Electromagnetic Linear Machines with Dual Halbach Array

Liang Yan · Lu Zhang · Juanjuan Peng
Lei Zhang · Zongxia Jiao

Electromagnetic Linear Machines with Dual Halbach Array

Design and Analysis

Springer

Liang Yan
Science and Electrical Engineering
Beihang University, School of Automation
Science and Electrical Engineering
Beijing
China

Lei Zhang
Science and Electrical Engineering
Beihang University, School of Automation
Science and Electrical Engineering
Beijing
China

Lu Zhang
Science and Electrical Engineering
Beihang University, School of Automation
Science and Electrical Engineering
Beijing
China

Zongxia Jiao
Science and Electrical Engineering
Beihang University, School of Automation
Science and Electrical Engineering
Beijing
China

Juanjuan Peng
Science and Electrical Engineering
Beihang University, School of Automation
Science and Electrical Engineering
Beijing
China

ISBN 978-981-10-2307-1 ISBN 978-981-10-2309-5 (eBook)
DOI 10.1007/978-981-10-2309-5

Library of Congress Control Number: 2016947739

Printed on acid-free paper

This Springer imprint is published by Springer Nature
The registered company is Springer Science+Business Media Singapore Pte Ltd.

The book is dedicated to authors' families

Foreword

This monograph is a very good reference for linear machine analysis to engineering students, academic researchers and practicing industrial engineers alike. It provides a unique insight into the operation of linear machines and presents some innovative, very high performance machines. A linear machine is an electromagnetic device that can directly produce linear force and translational motion along its length. Industry applications typically adopt hydraulic, pneumatic, or mechanical devices in the form of roller screws or ball screws to achieve linear actuation. Linear electrical machines can however be a very attractive alternative due to their potential higher efficiency, faster response, and lack of physical contact between the stationary and moving element. One drawback which often limits its adoption is the maximum force a linear machine can produce for a given weight or volume when compared to mechanical alternatives. This is not an issue in linear propulsion applications such as trains and amusement rides or in precision motion control applications but is a major issue when relatively high forces are required in applications such as aircraft control surface actuators. This monograph presents effective machine concepts which can significantly increase the net force output for a given weight (force density) of linear machines. Details of construction aspects and experimental results are presented.

This monograph elegantly extends the conventional two-dimensional (2D) arrangement of magnet arrays to 3D space patterns. This leads to novel concepts with opportunities to improve the achievable output force density. A novel dual Halbach array is proposed to achieve two tangible advantages. One to increase the radial magnetic flux density, and thus the force output of the system, and the other to reduce the axial magnetic flux density, and thus depresses significantly the vibration and noise. Impressive output performance can be achieved with this new topology and design methodology. This methodology can also be applied to rotary machines.

There are many books introducing the operating principles, modeling, design and control algorithms of electric machines. However, these books seldom concentrate on the design optimization and consequent analysis of output performance.

This is especially the case for systems with novel structures. This monograph is unique in that it is complementary to previously published books but also covers detailed formulation of 3D magnetic fields, mathematical modeling of output force, inductance derivation, design optimization and aspects of practical development and experimental performance validation. This monograph is a good reference for practitioners and professional experts in the field of electric machines. The innovative design concepts presented will surely spawn further creative designs of electrical machine structures.

Nottingham, UK
May 2016

Christopher Gerada

Preface

A linear machine is a counterpart of rotary machine except that the stator and rotor are unrolled, so that instead of producing torque and rotation, it generates force and translation along its length. Typically, there are two ways to improve the force output of linear machines, i.e., increasing current input and increasing magnetic flux density in the air gap. Although large current input is a straight way to improve the force output, it may lead to high temperature and cause the problem of heat dissipation. The thermal property of materials strictly constrains the performance improvement of electric machines. Employment of large size of permanent magnets (PMs) also helps to increase the flux density and force output. However, the system size is increased in an exponential way. The study in this monograph aims to increasing magnetic flux density and force output by reorganizing the magnet poles. Specifically, the conventional two-dimensional (2D) magnet arrangement is extended to magnet array in 3D space, and thus a novel dual Halbach array is proposed for the design of linear machines. It offers two advantages. One is to increase the magnetic flux density in radial direction, which in turn helps to improve the axial force output. The other is to decrease the flux density in axial direction, and thus depresses the vibration in radial direction. This monograph covers systematic study on design and analysis of tubular linear machines with dual Halbach array.

Chapter 1 introduces fundamental concepts, advantages, applications, and typical designs of linear machines. Different types of magnet arrays such as alternatively radial magnetization and alternatively axial magnetization are presented and analyzed. As a result, the dual Halbach array is proposed to increase the force output of the system. It is worth pointing out that the proposed dual Halbach array is applicable to other electric machines such as flat linear machines and rotary machines.

Chapter 2 presents the magnetic field formulation of the proposed linear machine. Magnetic field formulation is one precondition of force modeling. It helps to optimize the magnet arrangement and thus the system output performance. The governing equations are obtained based on magnetic properties of region materials in the linear machine. The general solution of magnetic vector potential is thus derived. By utilizing boundary conditions between different regions, the particular solution of magnetic vector potential is obtained. Finally, we have the magnetic flux density from the curl of magnetic vector potential. The magnetic flux density consists of three components in the space, one of which is equal to zero. Numerical results are employed to validate the derived magnetic field formulation.

Based on the magnetic field formulation, Chap. 3 mainly concerns the mathematical modeling of force output. Force output model is important for design optimization and control implementation of linear machines. Lorenz force law is especially suitable for force modeling of current-carrying conductors in external magnetic field, and thus it is used in this study. The force output for single-phase, double-phase, and three-phase windings are formulated. The force output of three-phase winding is also related to the starting point of mover. Therefore, the corresponding mathematic models are obtained. Similarly, numerical results are used to validate the analytical models of force output.

Chapter 4 focuses on analysis of armature reaction field and inductance. So far most studies of electric machines concern only the magnetic field of permanent magnet poles, whereas less work has been done on armature reaction field. Analysis of armature reaction field and inductance is extremely important for design and control implementation of electromagnetic machines. This chapter proposes a novel analytical modeling method to predict the armature reaction field of the coreless PM tubular linear machine. Unlike the conventional modeling approach, the proposed method formulates the armature reaction field for electromagnetic machines with finite length, so that the model precision can be improved. In addition, winding inductance is also analytically formulated to facilitate dynamic motion control based on the reaction field solutions. Numerical result is subsequently obtained with finite element method, and employed to validate the derived analytical models.

By utilizing the analytical models obtained in previous chapters, Chap. 5 studies the parameter optimization and back iron influence of tubular linear machines with dual Halbach array. Penalty method and scanning method are employed to conduct parameter optimization to maximize the force output for a given machine volume. The influence of structure parameters on force output and force ripple is discussed, and thus the optimal parameter values are determined. Ferromagnetic material is one key component that influences the force generation of electric machines significantly. Numerical computation from finite element method is employed to calculate the force output of linear machines with various types of back iron patterns including internal back iron, external back iron, double-sided back iron, and without back iron to achieve large force output in a certain volume.

Chapter 6 presents the experimental investigation on the tubular linear machine with dual Halbach array. One research prototype has been developed. An automatic measuring apparatus has also been developed to measure the magnetic field distribution in the space. The measuring procedure and corresponding date processing and analysis are presented. Similarly, measuring apparatuses for force output and armature reaction field are built up. The experimental results of the magnetic field distribution, the force output variation, and the armature reaction field are visualized, and utilized to verify the analytical models developed in previous chapters.

Chapter 7 concludes the major works in this monograph. More subsequent works could be conducted on the proposed novel design of linear machines. The same design method and analyzing approaches could be implemented to other linear and rotary machines.

The authors would like to acknowledge the financial support from the National Nature Science Foundation of China (NSFC) under the Grant No. 51575026, National Key Basic Research Program of China (973 Program, 2014CB046406), NSFC51235002, NSFC51175012 the Fundamental Research Funds for the Central University, and the Science and Technology on Aircraft Control Laboratory.

Beijing, China Liang Yan
May 2016

Contents

Acronyms

2/3D	Two/three-dimensional
AC	Alternating current
BCs	Boundary conditions
CM	Coenergy method
CNC	Computer numerical control
DC	Direct current
EALS	Electromagnetic aircraft launch system
EMC	Equivalent magnetic circuit
FE	Finite element
FEM	Finite element method
LFE	Longitude fringe effect
MMF	Magnetomotive force
MST	Maxwell stress tensor
PC	Personal computer
PM	Permanent magnet
PMLMs	Permanent magnet linear machines
PMTLM	Permanent magnet tubular linear machine
RMS	Root mean square

List of Figures

List of Tables

Chapter 1
Introduction

Abstract Permanent magnet (PM) linear machine generates linear motions directly without rotation-to-translation conversion mechanisms, which significantly simplifies system structure, and improves system working efficiency, dynamic response and control performance. In this chapter, the history of linear machines are briefly introduced. Then the fundamental concepts, advantages, and applications of linear machines are presented. Subsequently, typical magnet arrays in conventional linear machines are discussed. Following that, the research objective and scope of the tubular linear machine with dual Halbach array is proposed. Finally, the outline of this monograph is presented.

1.1 History of Linear Machine

The history of linear machine [1] can be traced back to 1840s. Charles Vheatstone at King's College in London proposed and developed a primary linear machine, although it was not very successful. A feasible linear induction motor is patented in 1905 by Alfred Zehden for driving trains or lifts. Thirty years later, a German engineer, Hermann Kemper, built a working model of linear machine. In the late 1940s, Eric Laithwaite at Manchester University and later Professor of Heavy Electrical Engineering at Imperial College in London, developed the first full-size linear machine. In the single sided version, the magnetic levitation is utilized. The magnetic repulsion force levitates the conductor away from the stator, and carries it along in the direction of moving magnetic field. In 1945, Westinghouse Company studied an aircraft catapult driven by a linear motor. And this study has gained much achievement. Since then, linear motors have received wide attention in practical applications, and stepped into new stage of research and development. The discovery of many new materials and the innovation of control technology promote the development and application of linear machines, for example, MHD pump and linear motor conveying device have been developed in industries. After 1970s, SIEMENS Company, Westinghouse Company and Panasonic Company have paid more attention on the investment of linear machines, and a large number of linear machines with different features have emerged in market.

L. Yan et al., *Electromagnetic Linear Machines with Dual Halbach Array*, DOI 10.1007/978-981-10-2309-5_1

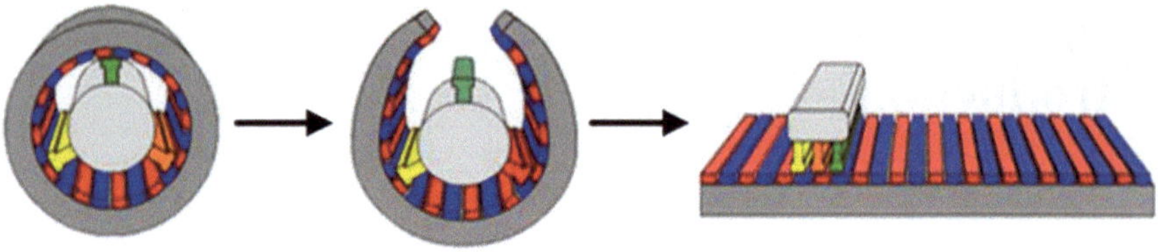

Fig. 1.1 Evolution of rotary motor to linear machine

1.2 Advantages of Linear Machine

An electric machine is a device that converts electricity into mechanical motions. The linear machine can be considered as one evolution of rotary machines in the structure as shown in Fig. 1.1. Permanent magnets (PMs) are mounted on the stator and windings are on the mover, or vice versa. The interaction between magnetic field of PM poles and current input in the windings generates force in the unrolled direction. Compared with conventional linear drive system consisting of rotary motors and ball screws, linear machines have many merits as follows:

- Linear machines are direct drive systems that do not need ball screws, cranks, gears, or any other motion conversion mechanisms, which helps to achieve compact size of system.
- The motion accuracy of linear motors can be improved significantly, because there is no backlash accumulation from screws. It could be utilized for the development of servo systems with high precision.
- The working efficiency of linear machines is relatively high as there is no energy loss by intermediate transmissions. In addition, linear machine can achieve magnetic suspension to decrease the friction and vibration between mover and stator, and thus increase the power efficiency.
- The mover can be connected to the payload directly. There is no additional inertia moment produced by linkages or ball screws. Therefore, the mass of moving components can be decreased, and thus the dynamic output performance of linear machines can be improved.
- Short primary structure is generally used in the design. Therefore, linear machines have efficient heat dissipation due to its open architecture.
- As there is no centrifugal force, the linear machine can achieve ultra high-speed translations.

1.3 Applications of Linear Machines

Because of the above-mentioned advantages, linear machines are widely used in military field [2], civilian field [3, 4], and other industrial systems.

1.3.1 Linear Machines in Military Field

The application of linear machines in military field started very early [5, 6], such as the aircraft catapult in [7]. Conventionally, the steam catapult devices are utilized in aircraft carriers. Large mass and big volume of the steam catapult lead to high operating costs. To increase the taking-off speed of aircrafts and reduce the operating cost, United States Navy (USN) starts to build the electromagnetic aircraft launch system (EMALS) with linear machines. Compared with the traditional steam catapult devices, EMALS can achieve high efficiency, and the exit speed of catapult is adjustable. At the early stage, because the energy storage devices and control technology were not very mature, the development of electromagnetic launch system was not successful. However, the exploration of linear machine's application in military field has never stopped. In recent years, the electromagnetic ejection technology based on linear machines has gained more attention. The electromagnetic catapult linear machine developed by NAVAIR Lakehurst is shown in Fig. 1.2.

1.3.2 Linear Machines in Civilian Field

Linear machines are also widely used in civilian field, such as washing machine, sewing machine, curtain machine, and so on. It is worth mentioning that linear compressors are utilized in air conditioners, refrigerators, and other civilian products

Fig. 1.2 Linear machine for electromagnetic catapult [8]

(a) The cross-belt sorting system

(b) The tilt-tray sorting system

Fig. 1.3 Linear machines for applications of logistics [11]

in recent years. Compared with traditional rotary compressors, linear compressors have the merits of high efficiency and low-noise. LG has successfully developed a linear oscillator compressor, and implemented it into refrigerators, although its structure is somehow complex.

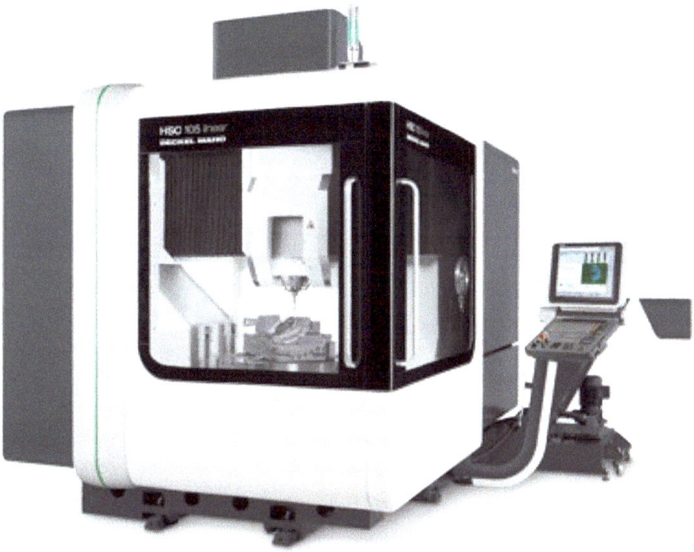

(a) High-precision machining center [14]

(b) Laser cutting center [15]

Fig. 1.4 Linear machines for applications of Industrial Automation

1.3.3 Linear Machines in Logistics

Conventionally, linear transmission systems consisting of rotary motors and ball screws are utilized in logistics, which unavoidably influences the working efficiency. To solve the problem, some companies have developed transmission systems and material handing systems driven by linear machines. For example, Swiss post

(a) MLX01-2 experimental maglev train in Japan [16]

(b) Maglev train in Shanghai [17]

Fig. 1.5 Linear machines for applications of rail transportation office developed the pouch hanging transmission equipments, and BEUMER company developed cross-belt sorting system and tilt-tray sorting system as shown in Figs. 1.3a, b, respectively. These systems are driven by linear machines, and thus can achieve advantages of high efficiency, low vibration and fast response.

1.3.4 Linear Machines in Industrial Automation

Linear machines have wide applications in industrial automation [9, 10], such as pumps in metallurgical industry, automatic plotters and punches. The linkages, gears and worms in conventional linear transmission mechanisms increase the system backlash more or less, and reduce the motion precision of end-effector. The employment of linear machines helps to improve the motion precision in such systems significantly. Figure 1.4a presents a model of computer numerical control (CNC) products from DMG company driven by linear machines. Figure 1.4b is a laser cutting center. The improvement of design and control technology of linear machine promotes its implementations in high performance machining systems.

1.3.5 Linear Machines in Rail Transportation

The linear machines can be implemented into rail transportation [12, 13]. The maglev train driven by electromagnetic force directly is one typical vehicle with magnetic levitation. It can achieve low friction, and thus very high speed compared with conventional trains. The application of linear machines in maglev rail transportation began in 1980s. It has gained preliminary success in Canada, Japan, China and other countries. Figure 1.5a shows the MLX01-2 experimental maglev train in Japan, and Fig. 1.5b shows the maglev train in Shanghai that adopts the high-speed maglev technology.

1.4 Magnet Patterns of Linear Machines

Magnet pattern influences the output performance of linear machines significantly. Different magnet patterns have been proposed by researchers for the design of linear machines. Wang et al. presents the design optimizations of an iron-cored tubular permanent magnet machine [18–20]. The permanent magnet poles are magnetized radially as shown in Fig. 1.6a. It has advantages of high efficiency and high power density, and the disadvantages of lower winding factor for the fundamental electromotive force. Analytical formulas are derived to predict the open-circuit electromotive force, the thrust force, the iron loss, and the winding resistance and inductances, as well as the converter losses. The same magnetization pattern can also

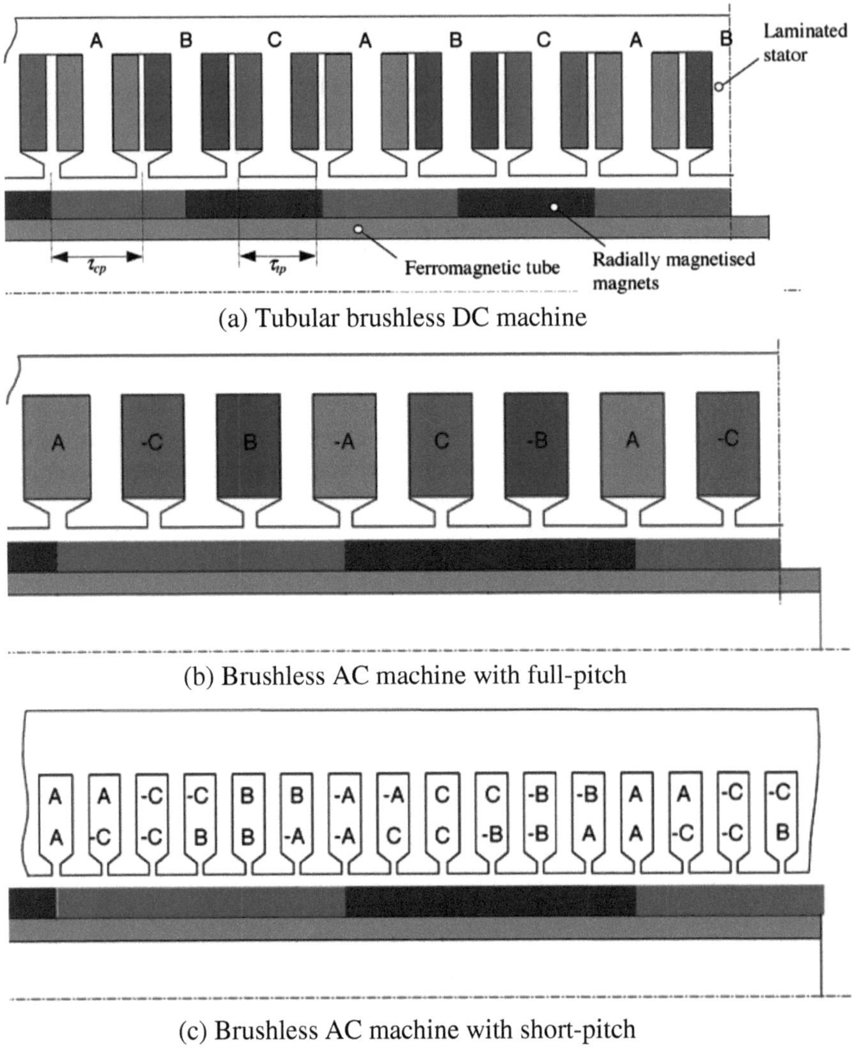

Fig. 1.6 Winding arrangements for tubular PM linear machines [19]

be implemented to machine design with different winding phases in Figs. 1.6b, c to reduce EMF harmonics and force ripple. They all have a integer ratio of slot number to pole number, leading to larger cogging force.

A double-sided slotted torus axial flux permanent magnet motor suitable for direct drive of electric vehicle is proposed in [21]. Its construction is illustrated in Fig. 1.7. The magnet poles are arranged alternatively along the rotor circumference, and magnetized in axial direction. The motor can be easily mounted compactly onto a vehicle wheel, fitting the wheel rim perfectly.

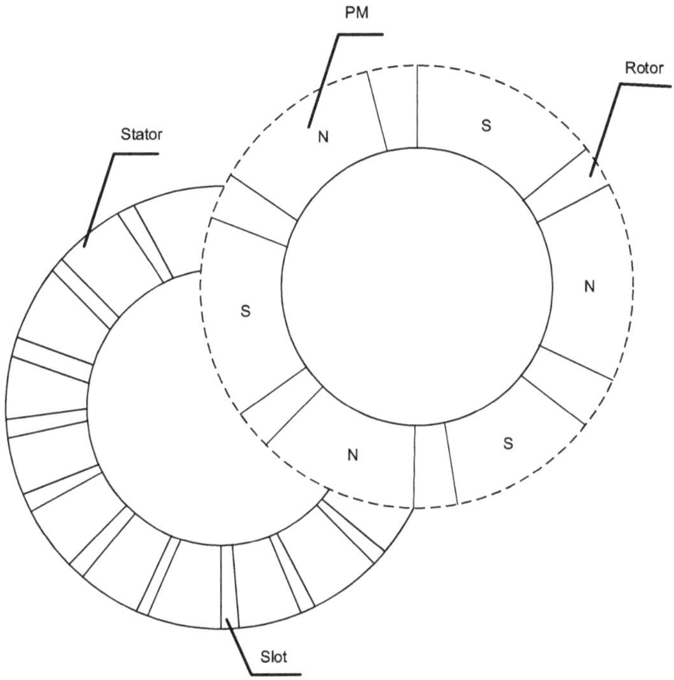

Fig. 1.7 An axial flux permanent magnet motor [21]

Nirei et al. presents the evaluation of a moving-coil-type cylindrical linear DC motor [22]. The structure is illustrated in Fig. 1.8. It consists of a coil, permanent magnets, yokes, a coil bobbin held by the arms of a coil holder and linear bearings. There are 16 pieces of permanent magnets, and they are fitted along the inner surface of the outer yoke and magnetized radially. The yokes are solid steel blocks. The inner yokes are separated into four pieces to provide space for the arms of the coil holder.

Kim et al. presents a tubular linear brushless permanent magnet motor as shown in Fig. 1.9 [23]. It has a slotless stator to provide smooth translation without cogging. In this design, the magnets in the moving part are oriented in an NS-NS SN-SN fashion which leads to higher magnetic force near the like-pole region.

Wang et al. describes the design and experimental characterisation of a reciprocating linear permanent magnet generator developed for on-board generation of electrical power fix telemetry vibration monitoring systems as shown in Fig. 1.10 [24, 25]. The two axially magnetised sintered NdFeB magnets and the mild-steel pole pieces give rise to an essentially radial magnetic field in the region occupied by the generator winding. The moving permanent magnet plunger is supported at each end by beryllium copper disc springs, which provide a high degree of radial stiffness, and an axial stiffness that can be accurately controlled by appropriate design of the spiral grooves.

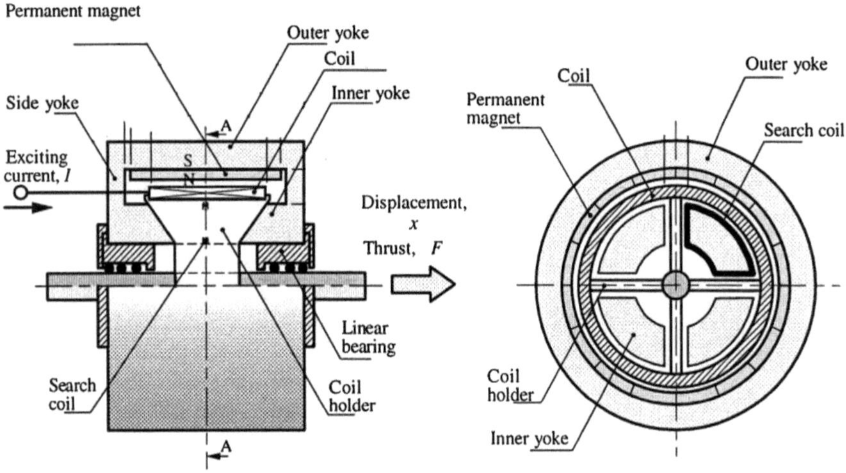

Fig. 1.8 Structure of the moving-coil-type linear DC motor [22]

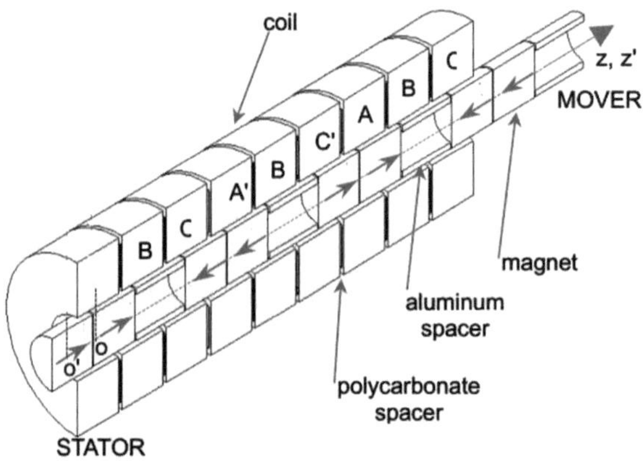

Fig. 1.9 Structure of the moving-coil-type linear DC motor [23]

Buren et al. presents the design and optimization of a linear electromagnetic generator suitable to supply power to body-worn sensor nodes [26]. The design is based on an air-cored tubular architecture and a flexible translator bearing. As shown in Fig. 1.11, the permanent magnets magnetized axially are mounted on the translator.

The influence of the leading design parameters on the performance of slotless tubular permanent magnet machines, and the relative merits of different tubular machine topologies, are deduced from analytical field solutions [27]. Linear machines with surface-mounted radially magnetized magnets and with axially magnetized magnets are analyzed in detail. Reference [28] describes an improved design for an axially magnetized tubular permanent magnet machine. Compared with a conventional axially magnetized tubular machine, it has a higher specific force capability and

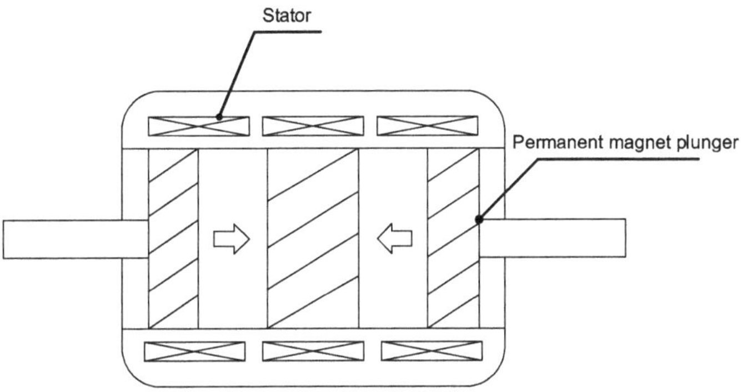

Fig. 1.10 Schematic of reciprocating linear generator system [24]

Fig. 1.11 A linear vibration-driven electromagnetic micro-power generator [26]

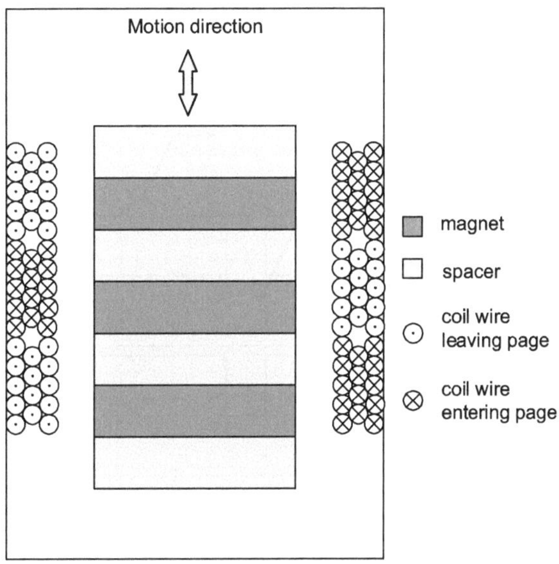

requires less permanent magnet material. As shown in Fig. 1.12, both the permanent magnets and the associated pole pieces are annular shaped and supported by a non-ferromagnetic rod. It eliminates the need for a nonmagnetic tube to contain the magnets and pole pieces, so that the effective magnetic air gap between the stator and armature is reduced. It also reduces the volume of permanent magnet material. Although there is some flux leakage via the central rod, the reduction in force capability can be relatively small.

The design and analysis of a brushless DC linear motor is introduced in [29]. Figure 1.13 shows the side section of magnets and coil in the air gap. The magnets are magnetized in the alternatively radial direction, and the current passes the area of coil for same direction.

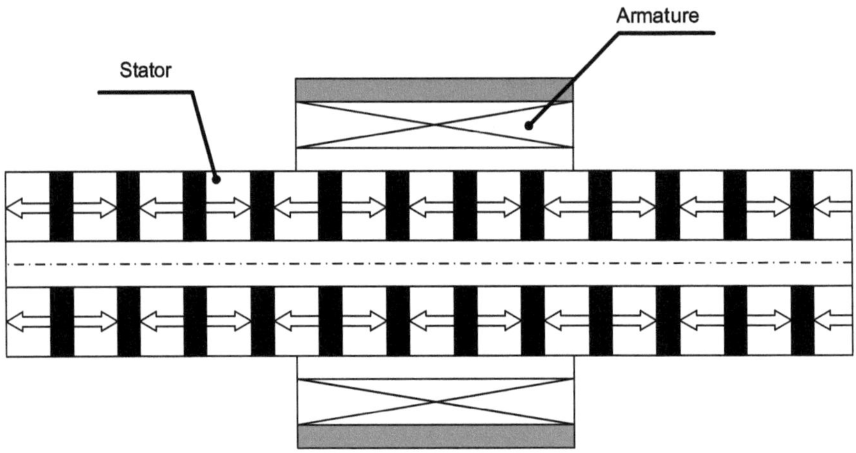

Fig. 1.12 An improved axially magnetized tubular PM machine topology [28]

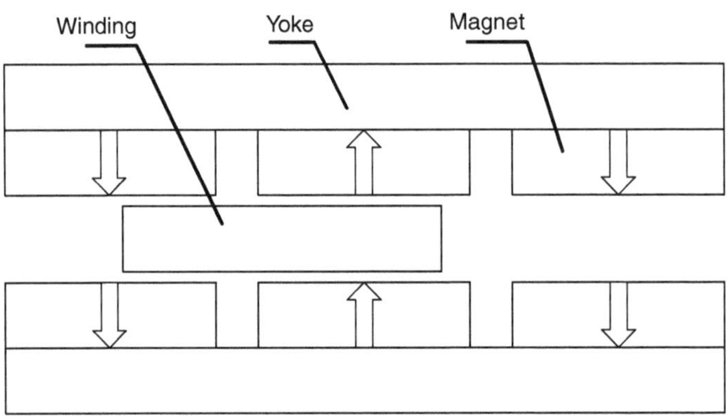

Fig. 1.13 Sectional view of a brushless DC linear motor [29]

1.5 Objective and Scope of the Study

The objective of this monograph is to extend the conventional two-dimensional (2D) magnet arrangement to 3D arrays, and thus propose the novel dual Halbach array. It helps to increase the radial component of magnetic flux density and thus improve the force output, and decrease the axial component of flux density and thus depress the vibration and noise of linear machines. The scope of this monograph is mainly on design and analysis of tubular linear machines with dual Halbach array, including concept design, formulation of magnetic field, mathematical modeling of force output, modeling of armature reaction field and inductance, design optimization, development of research prototype, and experimental investigation. Figure 1.14 illustrates the major tasks that have done in this study.

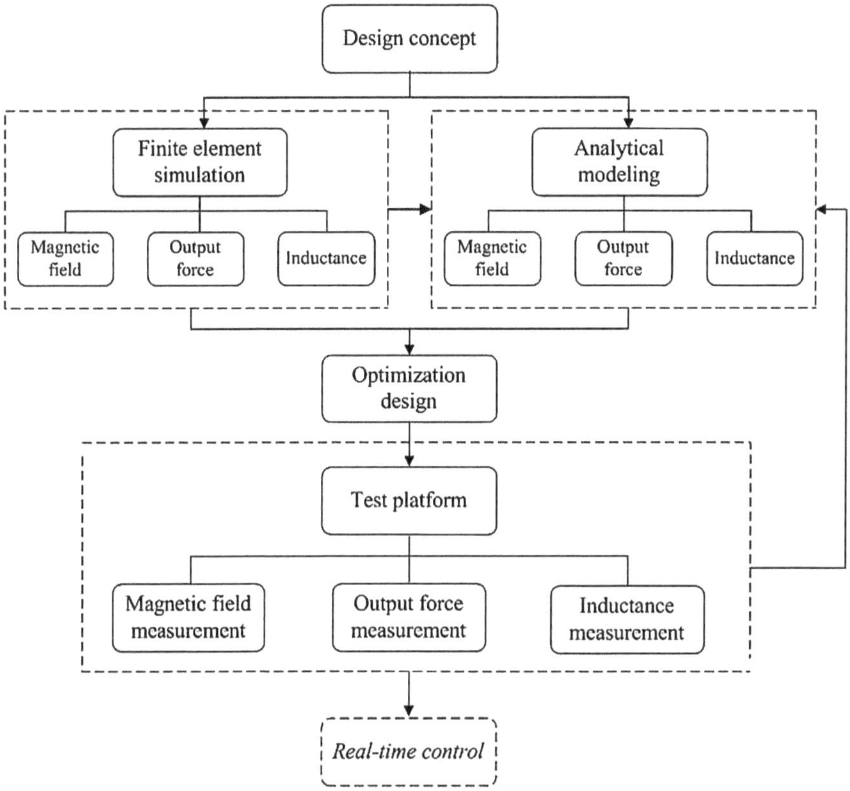

Fig. 1.14 Scope of the book

- **Concept design**: Concept design is the fundamental of subsequent studies. A variety of magnet arrays in PM linear machines are investigated, and the advantages and disadvantages of these magnet arrays are analyzed and summarized. Following that, a novel dual Halbach array is proposed to improve the output performance of linear machines. The design concept of dual Halbach array is presented and its features are discussed.
- **Formulation of magnetic field in 3D space**: Based on the source-free property of magnetic field and Maxwell equations, Laplace's equation and Poisson's equation are derived for regions in the linear motor. Following that, by using Bessel functions and the harmonic expansion of magnetization vector, the analytical solution of magnetic vector potential is obtained. From the curl of magnetic vector potential and boundary conditions, generic magnetic field model of linear machines in 3D space is achieved.
- **Mathematical modeling of force output**: By taking advantage of the magnetic field model and Lorentz force law, the force output for linear machines is formulated analytically. The analytical model of force output reflects the relationship

between geometric parameters and output performance, and thus is extremely important for the analysis of output performance and structure optimization.

- **Mathematical modeling of armature reaction field and inductance**: Analysis of armature reaction field helps to improve the dynamic control performance of linear machines. Fourier expansion and equivalent magnetic circuit are employed to formulate the magnetic field produced by the windings analytically. The total magnetic energy in 3D space is thus obtained. As a result, the mathematical model of the inductance can be formulated according to the relationship between the total magnetic energy and inductance.
- **Numerical computation and validation of analytical models**: Finite element method (FEM) has the advantage of high precision. It is employed to calculate the numerical results of magnetic field in 3D space, force output, and inductance of linear machines. The results are compared with the analytical models, which in turn validates the obtained mathematical models.
- **Design optimization of machine structure**: The obtained and validated analytical models are employed for structure optimization of the linear machine. The design optimization in this study includes parameter optimization and back iron design. The penalty method and scanning method are both utilized for the global optimization of major parameters. One set of optimal parameters are selected to improve the force output and minimize the force ripple. The analysis of back-iron patterns helps to improve the structure compactness and increase the axial force output.
- **Prototype development and experiments**: A research prototype of the linear machine with dual Halbach array is developed. Experimental apparatuses for measurement of magnetic field, force output, and inductance have been developed too. The experimental results are compared with analytical models, and validate the system operating principal.

All tasks indicated in Fig. 1.14 stem from the concept design. The analytical modeling is based on the concept design or magnet patterns of the novel structure, and it is extremely important for motion control of the linear actuator. The formulation of magnetic field is the prerequisite of force modeling. Numerical computation is employed to verify the developed analytical models. The force model is utilized to optimize the system structure. Based on the optimized structure, a research prototype and several experimental apparatuses are developed. Experimental investigation can be conducted on the magnetic field, force output and inductance to verify the mathematical models and operating principle of the proposed tubular linear machine. The theoretical and experimental results could be implemented into the real time motion control of the linear machine.

1.6 Book Organization

The remaining chapters of the monograph are arranged as follows. Chapter 2 introduces the working principle of the proposed linear machine with dual Halbach array. According to the PM pole arrangement, the magnetic field of the system is formulated. Chapter 3 models the output force of linear machine based on the magnetic field expression. Chapter 4 presents the armature reaction field and inductance of the tubular linear machine. By taking advantage of the magnetic field model and torque model, Chap. 5 presents the optimization design of the linear machine. In Chap. 6, a research prototype and several experimental apparatuses are developed to measure the flux density distribution, the output force variation and the reaction field. The data processing and presentation are discussed and the results are compared with theoretical models. Chapter 7 concludes the major works in this study.

References

1. Introduction of linear machine, https://en.wikipedia.org/wiki/Linear_motor
2. Kou B, Huang X, Wu H et al (2008) Thrust and thermal characteristics of electromagnetic launcher based on permanent magnet linear synchronous motors. IEEE Trans Mag 45(1):358–362, Jan 2009
3. Thornton R, Thompson MT, Perreault BM et al (2009) Linear motor powered transportation. Proc IEEE 97(11):1754–1757
4. Yan L et al (2009) The linear motor powered transportation development and application in China. Proc IEEE 97(11):1872–1880
5. Zhao M, Zou J, Xu Y et al (2011) The thrust characteristic investigation of transverse flux tubular linear machine for electromagnetic launcher. IEEE Trans Plasma Sci 39(3):925–930
6. Huang L, Yu H, Hu M et al (2013) Study on a long primary flux-switching permanent magnet linear motor for electromagnetic launch systems. IEEE Trans Plasma Sci 41(5):1138–1144
7. Aydemir MT, Zarko D, Lipo TA (2004) Design of a linear bulk superconductor magnet synchronous motor for electromagnetic aircraft launch systems. IEEE Trans Appl Supercond 14(1):54–62
8. Linear machine for aircraft lauch systems, http://insights.globalspec.com/article/1365/linear-motor-aircraft-launch-system-takes-the-steam-out-of-catapults
9. Shukor AZ, Fujimoto Y (2014) Direct-drive position control of a spiral motor as a monoarticular actuator. IEEE Trans Ind Electron 61(2):1063–1071
10. Commins PA, Moscrop JW, Cook CD (2011) Novel tooth design for a tubular linear motor for machine tool axis. In: Proceedings of the 2011 ieee intemational conference on mechatronics, pp. 660–665, Istanbul, Turkey, 13–15 Apr 2011
11. Cross belt sorting system, https://www.beumergroup.com/cn/home
12. Cao R, Cheng M, Mi CC et al (2014) Influence of leading design parameters on the force performance of a complementary and modular linear flux-switching permanent-magnet motor. IEEE Trans Ind Electron 61(5):2165–2175
13. Suzuki K, Kim YJ, Dohmeki H (2012) Driving method of permanent-magnet linear synchronous motor with the stationary discontinuous armature for long-distance transportation system. IEEE Trans Ind Electron 59(5):2227–2235
14. High-precision machining center, http://cn.dmgmori.com
15. Laser cutting center, http://www.hansme.com
16. Maglev train in Japan, http://hanyu.xdf.cn/201405/10002453.html

17. Maglev train in Shanghai, http://www.peixunwang.com.cn/hy/shgz/2015-10-14/167848.html
18. Wang J, Howe D, Jewell GW (2003) Fringing in tubular permanent-magnet machines: Part I. Magnetic field distribution, flux linkage, and thrust force. IEEE Trans Magn 39(6):3507–3516
19. Wang J, Howe D (2004) Design optimization of radially magnetized, iron-cored, tubular permanent-magnet machines and drive systems. IEEE Trans Magn 40(5):3262–3277
20. Wang J, Howe D (2005) Tubular modular permanent-magnet machines equipped with quasi-Halbach magnetized magnets-part I: magnetic field distribution, EMF, and thrust force. IEEE Trans Magn 41(9):2470–2478
21. Mahmoudi A, Rahim NA, Ping HW (2012) Axial-flux permanent-magnet motor design for electric vehicle direct drive using sizing equation and finite element analysis. Prog Electromagn Res 122:467–496
22. Nirei M, Tang Y, Mizuno T, Yamamoto H, Shibuya K, Yamada H (2000) Iron loss analysis of moving-coil-type linear DC motor. Sens Actuators Part A: Phys 81:305–308
23. Kim YH, Jin CS, Kim S et al (2002) Analysis of hybrid stepping motor using 3D equivalent magnetic circuit network method based on trapezoidal element. J Appl Phys 91(10):8311–8313
24. Wang J, Jewell GW, Howe D (1999) Design and experimental characterisation of a linear reciprocating generator. IEEE Proc Electr Power Appl 145(6):509–518
25. Wang J, Jewell GW, Howe D (1999) A general framework for the analysis and design of tubular linear permanent magnet machines. IEEE Trans Magn 35(3):1986–2000
26. Buren T, Troster G (2007) Design and optimization of a linear vibration-driven electromagnetic micro-power generator. Sens Actuators Part A: Phys 135:765–775
27. Wang J, Jewell GW, Howe D (2001) Design optimisation and comparison of tubular permanent magnet machine topologies. IEEE Proc Electr Power Appl 148(5):456–464
28. Wang J, Howe D, Jewell GW (2002) An improved axially magnetised tubular permanent magnet machine topology. IEEE Proc Power Electron Mach Drives 487:303–308
29. Hong JP, Kang DH, Joo SW, Hahn SC (2001) Variation of force density in bldc linear motor on the width of pm and coil. IEEE Int Symp Ind Electron Proc 979–981:2001

Chapter 2
Formulation of Magnetic Field

Abstract Permanent magnet array affects flux field distribution of electromagnetic linear machines significantly. A novel dual Halbach array is proposed in this chapter to enhance the flux density in air gap, and thus to improve the output performance of linear machines. Magnetic field in 3D space of a tubular linear machine with dual Halbach array is formulated based on Laplace's and Poison's equations. The analytical model is validated with numerical calculations from Ansoft. Furthermore, comparison of flux density between different permanent magnet arrays is done to highlight the advantages of dual Halbach array. The study shows that the employment of dual Halbach array does enhance the flux density and increase the force output of tubular linear machine. The research result in this chapter can also help the design optimization of electromagnetic linear machines.

2.1 Introduction

Electromagnetic linear machine generates linear motions directly without rotation-to-translation conversion mechanisms, which significantly simplifies system structure and improves system efficiency. It has wide applications in aeronautics [1, 2], transportation [3, 4], medical devices [5, 6] and so on. High flux density in linear machines is extremely important for applications requiring high force. The employment of permanent magnets offers electromagnetic linear machines a number of distinctive features [7] such as excellent servo characteristics and high power density. Optimized PM array is an effective way to improve flux density in linear machines. Axial and radial magnetization arrays are two of the most common arrangements of Permanent magnets. Magnet arrays with alternating magnetization directions are configured in [8] to produce radially directed flux density across the air gap of PM motors. A double-sided slotted torus axial-flux PM motor is designed for direct drive of electric vehicle in [9]. Reference [10] introduces the analysis and design of linear machines systematically. It provides a unified framework for several structure topologies. Halbach array is a promising magnet pattern due to its self-shielding property and sinusoidally distributed magnetic field in air space. It is widely applied in linear motor systems. For example, in [11], Halbach-arrayed PMs replace the

L. Yan et al., *Electromagnetic Linear Machines with Dual Halbach Array*, DOI 10.1007/978-981-10-2309-5_2

north-south magnet array in MagPie pipeline transportation system, and increase the motor propulsion force per Ampere significantly. In [12], PMs in Halbach array configuration are used for levitation, propulsion, and guidance of urban transportation systems, and achieve impressive performance. In this chapter, a tubular linear machine with dual Halbach array is proposed to further improve the magnetic flux density and thus the force output. It not only increases the radial component of flux density which is important for axial force generation, but also decreases the local force radial component which causes vibrations. Mathematical modeling of magnetic field is important for electromagnetic machines, as it could be used to predict output performance, such as flux linkage, field energy and force generation. Based on PM arrangement, magnetic field distribution in the machine is formulated with Laplace's and Poisson's equations analytically. Numerical result from finite element method is utilized to analyze and observe flux variation in three-dimensional space of the machine. The numerical result validates the analytical models. The obtained analytical model could be used for analysis of output performance and control implementation of electromagnetic linear machines with similar structures.

2.2 System Structure and Operating Principle

2.2.1 System Overall Structure

The schematic structure of the proposed PM tubular linear machine is illustrated in Fig. 2.1. The mover consists of winding phases mounted in the holds, whereas the stator is composed of dual Halbach magnet arrays that enclose the windings on internal and external sides.

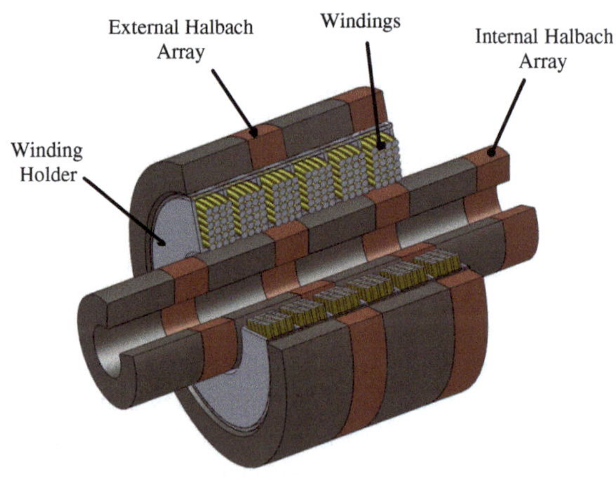

Fig. 2.1 Schematic drawing of linear machines with dual Halbach array

2.2.1.1 Stator Structure

The polarization pattern of each layer of the PMs is arranged in Halbach array. As indicated in Fig. 2.2, the Halbach array consists of alternatively magnetized PMs in radial directions separated with horizontally magnetized PMs. This array offers one impressive benefit, i.e., the flux density on one side of the PMs is significantly enhanced, whereas it is self-shielded on the other side. Figure 2.2 shows that the magnetization direction for horizontal magnets in both layers is opposite, while the magnetization for radial PMs is the same. This arrangement leads to the increase of radial component of flux density in between the two layers, and the reduction of axial components. Because only the radial component of flux density can create axial force on the mover, the dual Halbach array could improve the axial force output, and reduce the radial force disturbance of linear machines. The depression of axial components helps to reduce the deformation and vibration of linear machines.

2.2.1.2 Mover Structure

Movers in electrical linear machines can be classified into two types: slotted one and slotless one. The former can usually produce high force density but also yield large force ripple due to the cogging forces. Furthermore, energy loss may exist due to the eddy current produced in the soft iron. In contrast, a slotless rotor structure eliminates the tooth ripple cogging effect and thus improves the dynamic performance and servo characteristics of linear machines, at the cost of a slight down of force output. For some particular applications such as precision manufacturing and assembly, slotless mover has greater advantages than the slotted one. Therefore, in this study, we choose the slotless type for the mover design of the linear machines.

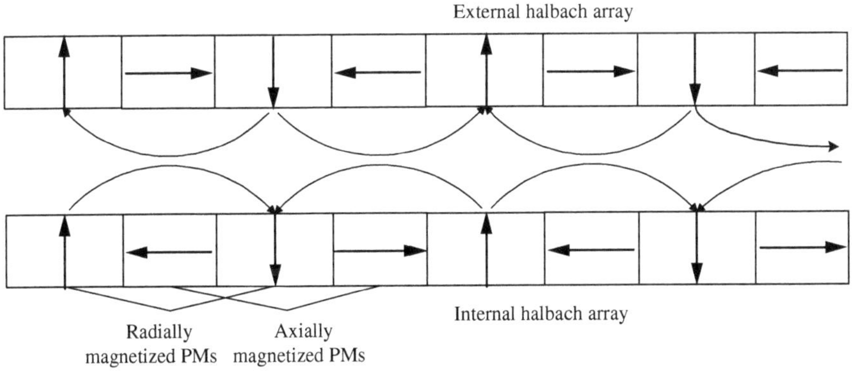

Fig. 2.2 Polarization pattern of dual Halbach array

2.2.2 Operating Principle

The operating principle of PM tubular linear machine with dual Halbach array is based on the Lorentz force law, i.e., force is generated by interaction between the magnetic field of PMs and the current-carrying conductors. The direction and magnitude of the force is determined by the cross product of current vector on segmental winding and magnetic flux density. Therefore, the radial component of flux density can produce forces on the mover in axial direction. The force created by the axial flux component and winding segments is symmetric around the machine axis, and thus vanishes in total. The stator and mover can be switched depending on what part we want to move for particular applications.

2.3 Governing Equations of Flux Field

Mathematical modeling of magnetic field is important for electromagnetic machines, as it could be used to predict output performance, such as flux linkage, field energy and force generation [13–15]. Generally, there are two typical ways to formulate magnetic field of electromagnetic machines, i.e., FEM [16] and magnetic equivalent circuits [17]. FEM is an efficient and accurate means to calculate magnetic field, taking full account of nonlinearity of iron material and induced currents in electrically conducting parts [18–20]. However, it is time consuming and cannot give much insight into design parameters [21]. Magnetic equivalent circuit can mainly be classified into lumped equivalent circuit and mesh-based one. The lumped equivalent circuit is the simplest approach to model the magnetic field, and allows to establish analytical relationships between design parameters and output performance. However, the technique suffers from inherent inaccuracy especially in presence of complex flux paths [22]. The mesh-based equivalent circuit is developed to achieve higher accuracy through a division of geometry [23]. However, it still requires some computational effort, although less than FEM, especially for complex models [24]. To obtain accurate knowledge of magnetic fields that directly relate motor geometry and output performance, a more sophisticated analytical field model that can compromise between accuracy and computation time is necessary [25]. Therefore, an analytical field model characterized by series expansions of the solution in terms of harmonic functions is established in this section. It will be validated by both finite element method and experiments. This approach could be applied to field modeling of other electromagnetic machines with similar magnet arrangements, and the derived parametric model is useful for analyzing the influence of parameters on output performance of electric machines.

2.3.1 Assumptions

The following assumptions are made to facilitate the subsequent study.

- The length of the linear machine is infinite, and thus the end effect of the magnetic field is ignorable.
- The magnetic permeability of back irons is infinite.
- Eddy currents in stator and mover are ignorable.

2.3.2 Magnetic Characterization of Materials

In formulation of the magnetic field, the machine space under study is divided into two regions based on magnetic characteristics. The air or coil space that has a relative permeability of 1.0 is denoted as Region 1. The permanent magnet volume filled with rare-earth magnetic material is denoted as Region 2. The back irons assumed infinite permeability is utilized to reduce magnetic energy loss, and enhance the flux density. The magnetic field property of Region 1 and 2 is characterized by the relationship between magnetic field intensity, **H** (in A/m), and flux density, **B** (in Tesla), as

$$\mathbf{B}_1 = \mu_0 \mathbf{H}_1, \tag{2.1}$$

$$\mathbf{B}_2 = \mu_0 \mu_r \mathbf{H}_2 + \mu_0 \mathbf{M}, \tag{2.2}$$

where μ_0 is the permeability of free space with a value of $4\pi \times 10^{-7}$H/m, μ_r the relative permeability of Permanent magnets, $\mathbf{M} = \mathbf{B}_{rem}/\mu_0$ the residual magnetization vector in A/m, and $\mathbf{B}_{rem}$ the remanence.

2.3.3 Governing Equations

The governing equations of magnetic field, i.e. Laplace's and Poisson's equations, are significant for the solution of magnetic field. It is known that magnetic field is a solenoid field or source-free field, i.e.,

$$\nabla \cdot \mathbf{B}_i = 0, \tag{2.3}$$

where $i = 1, 2$. It can be proved that for any vector, the divergence of its curl is always equal to zero. Thus, we can have a magnetic vector potential, $\mathbf{A}_i$, so that

$$\mathbf{B}_i = \nabla \times \mathbf{A}_i. \tag{2.4}$$

Because the curl of any function's (f) gradient is always equal to zero, we could have

$$\nabla \times \mathbf{A}_i = \nabla \times (\mathbf{A}_i + \nabla f),$$

which indicates that $\mathbf{A}_i$ may have multiple solutions. To uniquely determine its value, Coulomb gauge, $\nabla \cdot \mathbf{A}_i = 0$, applies as constraint. Under Coulomb gauge, we could have

$$\nabla \times \mathbf{B_i} = -\nabla^2 \mathbf{A}_i. \tag{2.5}$$

2.3.3.1 Laplace's Equation for Region 1

The combination of Maxwell's equations and Eq. (2.1) gives

$$\nabla \times \mathbf{B}_1 = \nabla \times \mu_0 \mathbf{H}_1 = \mu_0 \mathbf{J}. \tag{2.6}$$

Substituting Eq. (2.5) into (2.6) yields

$$\nabla^2 \mathbf{A}_1 = -\mu_0 \mathbf{J},$$

where $\mathbf{J}$ (A/m^2) is current density in the field. In this study, $\mathbf{J} = 0$. Therefore, the Laplace's equation for Region 1 is obtained as

$$\nabla^2 \mathbf{A}_1 = 0. \tag{2.7}$$

2.3.3.2 Poission's Equation for Region 2

The combination of Maxwell's equations and Eq. (2.2) gives

$$\nabla \times \mathbf{B}_2 = \mu_0 \mu_r \mathbf{J} + \mu_0 \nabla \times \mathbf{M}. \tag{2.8}$$

Similarly, Eqs. (2.5) and (2.8) yield the Poisson equation for Region 2

$$\nabla^2 \mathbf{A}_2 = -\mu_0 \nabla \times \mathbf{M}. \tag{2.9}$$

2.4 General Solutions to Magnetic Field

2.4.1 General Solution to Laplace's Equation

The magnetic field distribution of tubular linear machine is axially symmetric. Therefore, $\mathbf{A}_i$ has only one component, $A_{i,\theta}$. The Laplace equation in cylindrical coordinators can be simplified as

$$\frac{\partial^2 A_\theta}{\partial z^2} + \frac{\partial}{\partial r}\left(\frac{1}{r}\frac{\partial}{\partial r}(rA_\theta)\right) = 0. \tag{2.10}$$

Since A_θ is only a function of r and θ, we assume that

$$A_\theta = R(r)\,Z(z). \tag{2.11}$$

Substituting into Eq. (2.10) yields

$$\frac{1}{R(r)}\frac{\partial^2 R(r)}{\partial r^2} + \frac{1}{R(r)\,r}\frac{\partial R(r)}{\partial r} + \frac{1}{Z(z)}\frac{\partial^2 Z(z)}{\partial z^2} - \frac{1}{r^2} = 0, \tag{2.12}$$

while r and z are independent variables, and the third term as a function of z must be a constant. So the following formula is established

$$\frac{1}{Z(z)}\frac{\partial^2 Z(z)}{\partial z^2} = k^2. \tag{2.13}$$

Then Eq. (2.12) becomes

$$\frac{1}{R(r)}\frac{\partial^2 R(r)}{\partial r^2} + \frac{1}{R(r)\,r}\frac{\partial R(r)}{\partial r} + k^2 - \frac{1}{r^2} = 0. \tag{2.14}$$

Equation (2.13) can then be rewritten as

$$\frac{\partial^2 Z(z)}{\partial z^2} - k^2 Z(z) = 0. \tag{2.15}$$

Thus, Laplace's equation, Eq. (2.10), is separated into two equations, i.e., Eqs. (2.14) and (2.15). There are three possible solutions to Eqs. (2.14) and (2.15) according to the variation of k.

2.4.1.1 The First Solution

When $k^2 = 0$, the following equations are obtained

$$\begin{aligned} &Z(z) = E_0 + F_0 z, \\ &r^2\frac{\partial^2 R(r)}{\partial r^2} + r\frac{\partial R(r)}{\partial r} - R(r) = 0. \end{aligned} \tag{2.16}$$

The solution to Eq. (2.16) is

$$A_\theta = R(r)\,Z(z) = \left(C_0 r + D_0\frac{1}{r}\right)(E_0 + F_0 z). \tag{2.17}$$

However, as A_θ should be a periodic function of z, Eq. (2.17) is not the valid solution of A_θ.

2.4.1.2 The Second Solution

When $k^2 > 0$, the following equation is obtained as

$$\begin{aligned} Z(z) &= E_0 e^{kz} + F_0 e^{-kz}, \\ r^2 \frac{\partial^2 R(r)}{\partial r^2} &+ r \frac{\partial R(r)}{\partial r} + R(r)\left(k^2 r^2 - 1\right) = 0. \end{aligned} \tag{2.18}$$

Let $kr = x$. The second formula of Eq. (2.18) is

$$x^2 \frac{\partial^2 R}{\partial x^2} + x \frac{\partial R}{\partial x} + \left(x^2 - 1\right) R = 0.$$

The solution to Eq. (2.16) is

$$A_\theta = R(r)\,Z(z) = \left[C_0 J_1(kr) + D_0 Y_1(kr)\right]\left(E_0 e^{kz} + F_0 e^{-kz}\right). \tag{2.19}$$

Again, because Eq. (2.19) is not a periodic function of z, it is not the solution of Laplace's equation either.

2.4.1.3 The Third Solution

When $k^2 < 0$, the following equations are obtained

$$\begin{aligned} Z(z) &= B_0 \cos(mz) + B_1 \sin(mz), \\ r^2 \frac{\partial^2 R(r)}{\partial r^2} &+ r \frac{\partial R(r)}{\partial r} + R(r)\left(k^2 r^2 - 1\right) = 0. \end{aligned} \tag{2.20}$$

Let $mr = x$, Eq. (2.20) is

$$x^2 \frac{\partial^2 R}{\partial x^2} + x \frac{\partial R}{\partial x} - \left(x^2 + 1\right) R = 0.$$

The solution to Eq. (2.20) is

$$A_\theta = R(r)\,Z(z) = \left[C_0 I_1(mr) + D_0 K_1(mr)\right] * \left[E_0 \cos(mz) + F_0 \sin(mz)\right], \tag{2.21}$$

where m is a real number and defined with $k = jm$. In this case, A_θ is a periodic function of z. It is probably the solution of Laplace's equation. To determine the coefficients in above equation, constraints are needed. Because the axial component of flux density is antisymmetry on $z = 0$, we have $\mathrm{B_z}|_{z=0} = 0$. Substituting Eq. (2.21) into (2.4) gives

$$C_0 = 0,$$
$$D_0 = 0.$$

Therefore, the general solution to Laplace's equation is

$$A_\theta = [a_n I_1(mr) + b_n K_1(mr)] \sin(mz), \tag{2.22}$$

where $a_n = E_0$, $b_n = F_0$.

2.4.2 General Solution to Poisson's Equation

The Poisson equation in cylindrical coordinators is

$$\frac{\partial A_\theta}{\partial z^2} + \frac{\partial}{\partial r}\left(\frac{1}{r}\frac{\partial}{\partial r}(rA_\theta)\right) = -\mu_0 \nabla \times \mathbf{M}. \tag{2.23}$$

The general solution to the corresponding homogeneous equation of the Poisson's equation is

$$A_\theta = \sum_{n=1,2,\ldots}^{\infty} [a_n I_1(m_n r) + b_n K_1(m_n r)] \sin(m_n z). \tag{2.24}$$

To get the solution of Poisson equation, the right side of the equation needs to be substituted by harmonic expansion of magnetization vector. As illustrated in Fig. 2.2, the two Halbach arrays (Region 1 and Region 3) in the linear machine are composed of radial and axial magnets. The magnetization vector, $\mathbf{M}$, has two components M_r and M_z in r and z direction, respectively. It is given in cylindrical coordinators as

$$\mathbf{M} = M_r e_r + M_z e_z. \tag{2.25}$$

M_r is a non-continuous function with a period of $2\tau_p$, and its variation in z direction is shown in Fig. 2.3. To get the derivative of M_r, harmonic expansion is utilized,

$$M_r = \begin{cases} -M_0 & -\tau_p \le z < \frac{\tau_r}{2} - \tau_p \\ 0 & \frac{\tau_r}{2} - \tau_p \le z < -\frac{\tau_r}{2} \\ M_0 & -\frac{\tau_r}{2} \le z < \frac{\tau_r}{2} \\ 0 & \frac{\tau_r}{2} \le z < \tau_p - \frac{\tau_r}{2} \\ -M_0 & \tau_p - \frac{\tau_r}{2} \le z < \tau_p \end{cases},$$

where τ_r is width of radial magnets and τ_p is the pole pitch. It can be expanded with Fourier series as

$$M_r = \sum_{n=1,2\ldots}^{\infty} 4\left(B_{rem}/\mu_0\right) \frac{\sin\left[(2n-1)\frac{\pi}{2}\alpha_p\right]}{(2n-1)\pi} \cos(m_n z), \tag{2.26}$$

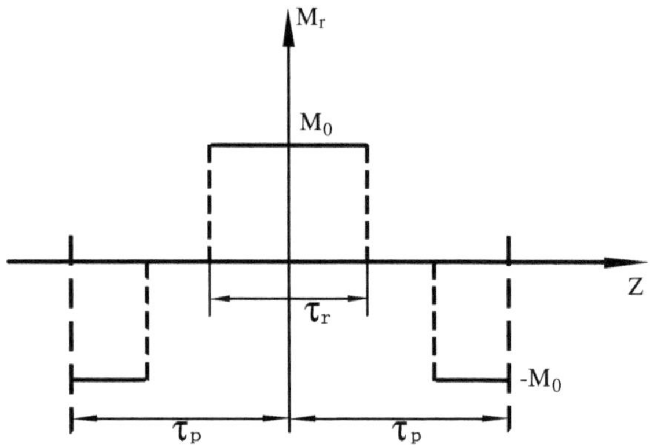

Fig. 2.3 M_r as a function of z

where $\alpha_p = \frac{\tau_r}{\tau_p}$, n is positive integer, and

$$m_n = {(2n-1)\pi}/{\tau_p}. \tag{2.27}$$

Therefore,

$$\begin{aligned} &-\mu_0 \nabla \times M \\ &= \sum_{n=1,2...}^{\infty} 4B_{rem} \frac{\sin\left[(2n-1)\frac{\pi}{2}\alpha_p\right]}{\tau_p} \sin(m_n z) \\ &= \sum_{n=1,2...}^{\infty} P_n \sin(m_n z), \end{aligned}$$

where $P_n = \frac{4}{\tau_p} B_{rem} \sin\left[(2n-1)\frac{\pi}{2}\alpha_p\right]$. The general solution of Poisson equation is obtained

$$A_{2\theta} = \sum_{n=1,2,...}^{\infty} \{[a_{2n} I_1(m_n r) + b_{2n} K_1(m_n r)] \sin(m_n z) + S(r,z)\}, \tag{2.28}$$

where $S(r,z) = R(r)Z(z)$ is a particular solution of Poisson equation. Substituting $S(r,z)$ into Poisson equation gives

$$\frac{1}{R(r)}\frac{\partial^2 R(r)}{\partial r^2} + \frac{1}{R(r)r}\frac{\partial R(r)}{\partial r} + \frac{1}{Z(z)}\frac{\partial^2 Z(z)}{\partial z^2} - \frac{1}{r^2} = R(r)Z(z)P_n \sin(m_n z). \tag{2.29}$$

Let $\frac{1}{Z(z)}\frac{\partial^2 Z(z)}{\partial z^2} == -{m_n}^2$. It is simplified as

$$Z(z) = \sin(m_n z), \tag{2.30}$$

$$r^2 \frac{\partial^2 R(r)}{\partial r^2} + r\frac{\partial R(r)}{\partial r} - R(r)\left({m_n}^2 r^2 + 1\right) = r^2 P_n. \tag{2.31}$$

Let $m_n r = x$. Equation (2.31) is reduced to

$$x^2 \frac{\partial^2 R(r)}{\partial x^2} + x \frac{\partial R(r)}{\partial x} - R(r)\left(x^2+1\right) = \frac{2x^2}{\pi} \frac{\pi P_n}{2{m_n}^2}.$$

Let $y = R(r) \frac{2{m_n}^2}{\pi P_n}$. The above equation becomes

$$x^2 \frac{\partial^2 y}{\partial x^2} + x \frac{\partial y}{\partial x} - y\left(x^2+1\right) = \frac{2x^2}{\pi}.$$

The above equation is a modified Struve equation. Its solution is

$$R(r) = \frac{\pi\, StruveL_1(m_n r)}{2{m_n}^2} P_n.$$

Therefore, the particular solution to Poisson equation is

$$S(r,z) = R(r)\, Z(z) = \frac{\pi\, StruveL_1(m_n r)}{2{m_n}^2} P_n \sin(m_n z), \tag{2.32}$$

where L_1 is the modified Struve functions [26] and appears as special solutions of inhomogeneous Bessel equations. As a result, the general solutions to Laplace and Poisson equations are

$$\begin{aligned} A_{1\theta} &= \sum_{n=1,2,\ldots}^{\infty} \left[a_{1n} I_1(m_n r) + b_{1n} K_1(m_n r)\right] \sin(m_n z), \\ A_{2\theta} &= \sum_{n=1,2,\ldots}^{\infty} \{\left[a_{2n} I_1(m_n r) + b_{2n} K_1(m_n r)\right] \sin(m_n z) \\ &\quad + \frac{1}{2} \frac{\pi\, StruveL_1(m_n r)}{{m_n}^2} P_n \sin(m_n z)\}. \end{aligned} \tag{2.33}$$

2.4.3 Solutions to Flux Density Distribution

From Eqs. (2.4), and (2.33), the general solutions of flux density is obtained

$$\begin{aligned} B_{r1} &= \sum_{n=1,2,\ldots}^{\infty} - m_n \left[a_{1n} I_1(m_n r) + b_{1n} K_1(m_n r)\right] \cos(m_n z), \\ B_{r2} &= \sum_{n=1,2,\ldots}^{\infty} - m_n \{\left[a_{2n} I_1(m_n r) + b_{2n} K_1(m_n r)\right] \cos(m_n z) \end{aligned}$$

$$+\frac{1}{2}\frac{\pi StruveL_1(m_n r)}{m_n{}^2}P_n\cos(m_n z)\}, \tag{2.34}$$

$$B_{z1} = \sum_{n=1,2,\ldots}^{\infty} m_n\left[a_{1n}I_0(m_n r) - b_{1n}K_0(m_n r)\right]\sin(m_n z),$$

$$B_{z2} = \sum_{n=1,2,\ldots}^{\infty} m_n\{\left[a_{2n}I_0(m_n r) - b_{2n}K_0(m_n r)\right]\sin(m_n z)$$

$$+\frac{1}{2}\frac{\pi StruveL_0(m_n r)}{m_n{}^2}P_n\sin(m_n z)\}.$$

B_{r1} and B_{z1} represents the radial and axial magnetic field in the winding region, while B_{r2}^{p} and B_{z2}^{p} are in magnet regions. The upper script, $p = 1, 2$, represents external and internal PMs respectively.

2.4.4 Boundary Conditions

Boundary conditions are necessary to determine the specific solutions of magnetic field. In this study, boundary conditions are employed to calculate the coefficients in the expression of magnetic flux density, such as a_{in}, b_{in}. The magnetic field follows certain rules along the boundary of two different media. For example, the flux density component perpendicular to the boundary is continuous in two neighboring media, and the tangential component of magnetic intensity is discontinuous by the amount of surface current at the boundary. In this study, surface current is zero. Therefore, the tangential component of magnetic intensity is continuous. The boundary conditions are

$$\begin{aligned} B_{1z}|_{r=R_s} &= \mu_0 M_{z1};\ B_{3z}|_{r=R_r} = \mu_0 M_{z3}, \\ B_{1r}|_{r=R_b} &= B_{2r}|_{r=R_b};\ H_{1z}|_{r=R_b} = H_{2z}|_{r=R_b}, \\ B_{2r}|_{r=R_a} &= B_{3r}|_{r=Ra};\ H_{2z}|_{r=R_a} = H_{3z}|_{r=R_a}. \end{aligned} \tag{2.35}$$

From the boundary conditions, the coefficients in magnetic flux density can be obtained.

2.5 Finite Element Analysis and Results

2.5.1 Overall Field Distribution

Figure 2.4 describes overall magnetic field distribution of a tubular linear machine with dual Halbach array. The major parameters used in the computation are given in Table 2.1. The magnets are sintered NdFeB with $B_{rem} = 1.2$(T) and $\mu_r = 1.0997$.

Fig. 2.4 Magnetic field distribution of machines with dual Halbach array

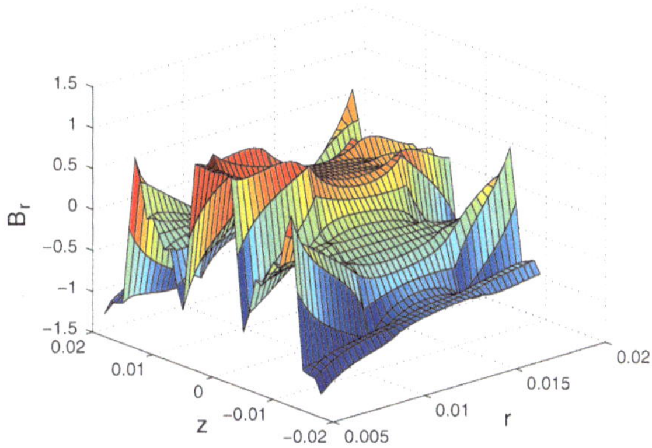

(a) B_r distribution

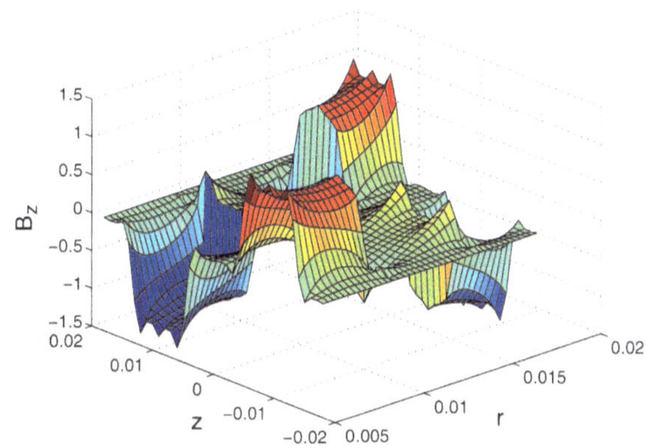

(b) B_z distribution

Table 2.1 Structure parameters in the numerical computations

Outer radius of external Halbach R_s	17 (mm)
Inner radius of external Halbach array R_b	14 (mm)
Outer radius of internal Halbach array R_a	9 (mm)
Inner radius of internal Halbach array R_r	5 (mm)
Pole pitch τ_p	9 (mm)
Axial width of radially magnetized PMs τ_r	18 (mm)

Magnetic field of a tubular linear machine is calculated with radial field B_r and axial component B_z by utilizing established analytical model. Figure 2.4a gives the variation of B_r, and Fig. 2.4b gives the variation of B_z in both magnet and winding regions. It is found that radial field is continuous in the radial direction while axial component jumps at the radial boundary of different materials. B_r and B_z variation is consistent with the distribution of remanence M_r and M_z in the magnet region. The magnetic field in the winding region tends to be uniform. Analytical model is then validated with finite element result according to two regions denoted by materials: magnet region and winding region. As it is assumed that the machine length is infinite, the finite element solutions are obtained by applying a master–slave boundary at the axial boundaries $z = 0$, $z = \tau_p$ and imposing symmetry boundary at $z = 0$.

2.5.2 Magnetic Field Variation in the Magnet Region

Figures 2.5 and 2.6 show magnetic field variation versus axial distance z at the center radius of internal and external Halbach arrays, i.e., $r = (R_r + R_a)/2$ and $r = (R_b + R_s)/2$, respectively. Magnetic field in either magnet region varies in line with

Fig. 2.5 Magnetic field variation at the center of internal magnet area

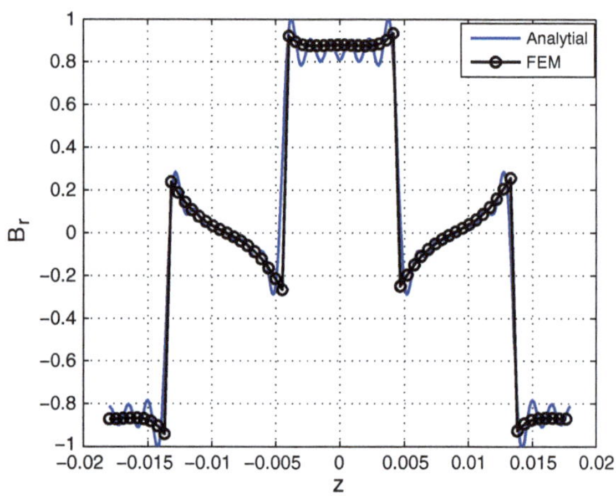

(a) B_r variation

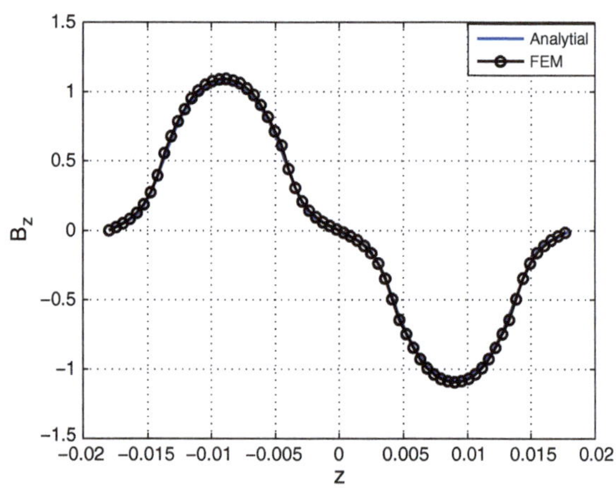

(b) B_z variation

Fig. 2.6 Magnetic field variation at the center of external magnet area

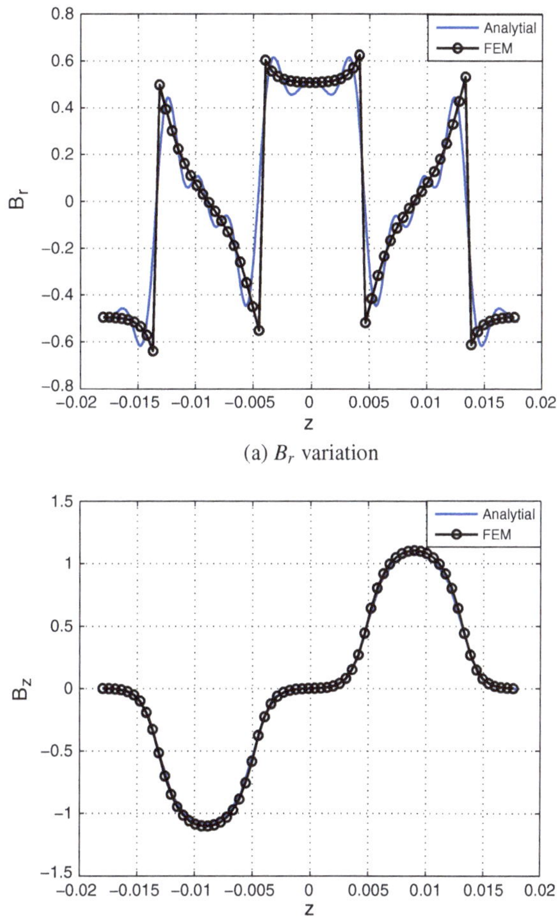

the magnetization vector M. Therefore, the radial flux component is even-symmetric about $z = 0$ mm, while the axial field is odd-symmetric. However, the radial flux density in the internal magnet area is greater than that in the external magnet area due to a decreasing section crossed by constant flux lines. The analytical results fit with the finite element results well. The difference is mainly caused by the simplification of models and FEM meshing.

2.5.3 Magnetic Field Variation in the Winding Region

Figure 2.7 shows B_r and B_z variation versus z at a constant radius $r = 12$ mm. The radial field is severely uniform in the radial magnet region, while the axial field

Fig. 2.7 Magnetic field variation in the winding region

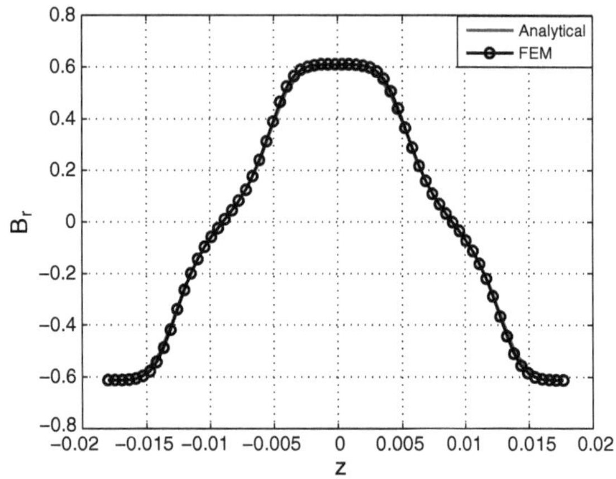

(a) B_r variation

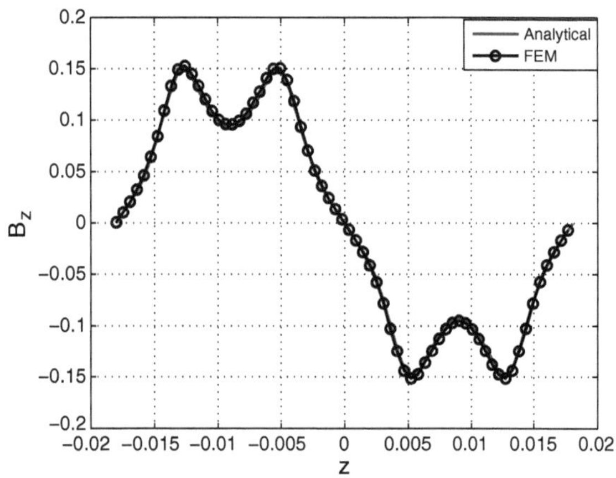

(b) B_z variation

reaches peak point at the axial boundary of magnets. The radial field variation related to force output is an approximate sine waveform which makes the motion control easy. It is found that the analytical results fit well with finite element results.

The foregoing study validates the analytical model with numerical calculations in Ansoft environment. The analytical model of an infinite length motor is proved to be accurate. However, longitude fringe effect (LFE) always exists in a tubular linear machine in practice because of finite length and it is uncertain whether the established motor fit well with a practical machine measurements.

2.6 Conclusion

A novel dual Halbach array is proposed in this chapter for the development of tubular linear machines. It helps to improve the radial flux, and reduce the axial flux. The 3D magnetic field distribution is formulated analytically based on Laplace's and Poisson's equations. Numerical computation of magnetic field is conducted with FEM method. It shows that the analytical model fits with the numerical result closely. The analytical model in this chapter can be used for design optimization and control implementation of tubular electromagnetic linear machines.

References

1. Aydemir MT, Zarko D, Lipo TA (2004) Design of a linear bulk superconductor magnet synchronous motor for electromagnetic aircraft launch systems. IEEE Trans Appl Supercond 14(1):54–62
2. Kou B, Huang X, Wu H et al (2009) Thrust and thermal characteristics of electromagnetic launcher based on permanent magnet linear synchronous motors. IEEE Trans Magn 45(1):358–362
3. Thornton R, Thompson MT, Perreault BM et al (2009) Linear motor powered transportation. Proc IEEE 97(11):1754–1757
4. Yan L (2009) The linear motor powered transportation development and application in China. Proc IEEE 97(11):1872–1880
5. Yamada H, Yamaguchi M, Karita M et al (1994) Acute animal experiment using a linear motor-driven total artificial heart. IEEE Transl J Magn Jpn 9(6):90–97
6. Yamada H, Yamaguchi M, Kobayashi K et al (1995) Development and test of a linear motor-driven total artificial heart. IEEE Eng Med Biol Mag 14(1):84–90
7. Mohammadpour A, Gandhi A, Parsa L (2012) Winding factor calculation for analysis of back EMF waveform in air-core permanent magnet linear synchronous motors. IET Electr Power Appl 6(5):253–259
8. Wang J, Howe D (2004) Design optimization of radially magnetized, iron-cored, tubular permanent-magnet machines and drive systems. IEEE Trans Magn 40(5):3262–3277
9. Mahmoudi A, Rahim NA, Ping HW (2012) Axial-flux permanent-magnet motor design for electric vehicle direct drive using sizing equation and finite element analysis. Prog Electromagn Res 122:467–496
10. Wang J, Jewell GW, Howe D (1999) A general framework for the analysis and design of tubular linear permanent magnet machines. IEEE Trans Magn 35(3):1986–2000
11. Fang J, Montgomery DB, Roderick L (2009) A novel magpipe pipeline transportation system using linear motor drives. Proc IEEE 97(11):1848–1855
12. Gurol HS (2009) General atomics linear motor applications: moving towards deployment. Proc IEEE 97(11):1864–1871
13. Torkaman H, Afjei E (2009) Magnetostatic field analysis regarding the effects of dynamic eccentricity in switched reluctance motor. Prog Electromagn Res M 8:163–180
14. Torkaman H, Afjei E (2010) Comparison of two types of dual layer generator in field assisted mode utilizing 3D-FEM and experimental verification. Prog Electromagn Res B 23:293–309
15. Torkaman H, Afjei E (2012) Comparison of three novel types of two-phase switched reluctance motors using finite element method. Prog Electromagn Res 125:151–164
16. Jian L, Chau KT (2010) Design and analysis of a magnetic-geared electronic-continuously variable transmission system using finite element method. Prog Electromagn Res 107:47–61

17. Touati S, Ibtiouen R, Touhami O et al (2011) Experimental investigation and optimization of permanent magnet motor based on coupling boundary element method with permeances network. Prog Electromagn Res 111:71–90
18. Lecointe JP, Cassoret B, Brudny JF (2011) Distinction of toothing and saturation effects on magnetic noise of induction motors. Prog Electromagn Res 112:125–137
19. Zhao W, Cheng M, Cao R et al (2012) Experimental comparison of remedial single-channel operations for redundant flux-switching permanent-magnet motor drive. Prog Electromagn Res 123:189–204
20. Mahmoudi A, Kahourzade S, Rahim NA et al (2012) Improvement to performance of solid-rotor-ringed line-start axial-flux permanent-magnet motor. Prog Electromagn Res 124:383–404
21. Musolino A, Rizzo R, Tripodi E (2012) Tubular linear induction machine as a fast actuator: analysis and design criteria. Prog Electromagn Res 132:603–619
22. Matyas AR, Biro KA, Fodorean D (2012) Multi-phase synchronous motor solution for steering applications. Prog Electromagn Res 131:63–80
23. Kim YH, Jin CS, Kim S et al (2002) Analysis of hybrid stepping motor using 3D equivalent magnetic circuit network method based on trapezoidal element. J Appl Phys 91(10):8311–8313
24. Amrhein M, Krein PT (2010) Induction machine modeling approach based on 3-D magnetic equivalent circuit framework. IEEE Trans Energy Convers 25(2):339–347
25. Liu C, Chau KT (2012) Electromagnetic design and analysis of double-rotor flux-modulated permanent-magnet machines. Prog Electromagn Res 131:81–97
26. The wolfram functions site, http://functions.wolfram.com/Bessel-TypeFunctions/StruveL/introductions/Struves/01/

Chapter 3
Mathematical Modeling and Analysis of Force Output

Abstract Force output is extremely important for electromagnetic linear machines. The purpose of this chapter is to formulate the force output of tubular linear machine with dual Halbach array. Based on the magnetic field model derived in the last chapter and Lorentz force law, the thrusts for single-phase, double-phase, and three-phase tubular linear machines are obtained analytically. The influence of the starting positions of the mover on the force output is also analyzed comprehensively. Numerical computation is conducted on the force output to validate the analytical solutions. The comparison shows that the analytical model fits with the numerical result well.

3.1 Introduction

The force modeling of electromagnetic linear machine is to establish the relationship between current input and force output mathematically. It can achieve two functionalities [1]. One typical usage is for computer simulation of the linear machine. Given one set of current inputs, force output of the linear machine can be calculated according to the force model. Based on forward dynamics of the machine, we can simulate how the machine moves under the action of the force. Another significant use of the force model is for position and velocity servo control. According to inverse dynamics of a linear machine, the desired force output can be computed from the desired displacement, velocity and acceleration of the mover. Then by utilizing the toque model, required currents can be calculated from the desired force output. These currents are supplied to the electric machine so that it can follow the desired motion.

Generally speaking, there are three types of methods that can be employed for the calculation of the force modeling of electromagnetic linear machine, i.e., Maxwell stress tensor method (MST), coenergy method and Lorentz force law. The MST approach computes local stress at all points of a boundary surface of a body, then sums the local stresses to obtain the global magnetic force/torque [2]. This classic method is usually used in finite element (FE) solution of force/torque computation [3]. For example, Adamiak et al. took advantage of MST to compute the thrust and normal force of a single-sided linear induction motor [4]. A method based on MST has been proposed by Zhou et al. to obtain the force ripples of a Halbach PM

L. Yan et al., *Electromagnetic Linear Machines with Dual Halbach Array*, DOI 10.1007/978-981-10-2309-5_3

ironless linear motor with trapezoid windings [5]. MST is convenient to determine the electromagnetic force on the interface and it can be used to describe the thrust conveniently. This method is very sensitive to the discrete density and contour shape. And it is difficult to define the boundary surface in three-dimensional problems.

The coenergy method [6] can be employed to obtain the force model of electromagnetic products analytically. Many electromagnetic products possess coils and ferromagnetic materials such as soft iron disrupted by narrow air gaps. By energizing the coils, flux lines are generated and flow through the soft iron and air gaps, forming a closed magnetic flux loop. The reluctance of air gap is much larger than that of soft iron. Therefore, the magnetic energy is mostly stored in the narrow air gaps. In this case, the magnetic energy can be formulated easily. This approach was utilized by Cho et al. to drive the thrust model of a hybrid type double-sided linear pulse motor [7]. Bazghaleh et al. has also obtained the traction and normal forces of a single-sided linear induction motor by using coenergy principle [8]. Coenergy method provide convenience for the calculation of the energy in the air gap. However, for the existence of hollow coils and large air gaps, this method is no longer applicable to the high-precision force modeling.

Alternatively, Lorentz force law can be adopted because energy formation is not required. It is very effective to calculate the thrust generated by energized conductor in external magnetic field. Lorentz force law [9] states that the force $d\mathbf{F}$ exerted on differential length segment $d\mathbf{l}$ of current-carrying conductor by external magnetic field $\mathbf{B}$ is equal to the multiplication of the current I_w and the cross product of $d\mathbf{l}$ and $\mathbf{B}$, i.e. $d\mathbf{F} = I_w d\mathbf{l} \times \mathbf{B}$. By using Lorentz force law, Wang et al. calculated the force of a class of tubular linear permanent magnet machines [10–13]. Another implementation of this method is found in the thrust formulation of a tubular oscillating permanent magnet actuator by Zhu et al. [14, 15].

Halbach array is a promising magnet pattern due to its self-shielding property and the ability to create relatively strong multi-pole, sinusoidally distributed magnetic field in the internal air space of electromagnetic linear machines. Based on the novel design of tubular linear machine with dual Halbach PM array proposed and the mathematic model of the magnetic field from magnetic vector potential in the last chapter, the per-phase force output of the system is obtained in this chapter. Moreover, the total thrust force and force ripple are formulated analytically for different phase patterns and winding arrangements. The derived analytical models are validated with numerical results. Based on the force models, the influence of structure parameters on force output and force ripple of a three-phase linear machines has been investigated. The study provides an effective tool for comprehensively understanding the operating principle and output performance of linear machines with dual Halbach arrays.

3.2 Formulation of Force Generation

3.2.1 Assumptions

Following assumptions are made to facilitate the subsequent study.

- The length of tubular linear machine is infinite, and thus the end effect of the magnetic field is ignorable.
- The permeability of iron parts is assumed infinite, and thus the saturation effect and hysteresis loss are ignorable. However, the saturation effect can also be taken into consideration by iteratively modifying the air gap length as explained in [16].
- Eddy currents in the stator and the mover are ignorable.

3.2.2 Thrust for Single-Phase Winding Pattern

To facilitate the subsequent discussion, we briefly recall the solution of flux density. The magnetic flux generated by PM arrays in tubular linear machine is symmetrically distributed around the machine axis. Therefore, there are only two components of the flux density. By taking advantage of magnetic vector potential and the boundary conditions in between different materials, the flux density distribution in the three-dimensional space can be represented as a function of r and z,

$$B_r = \sum_{n=1}^{\infty} -m_n \left[a_{3n} I_1 \left(m_n r\right) + b_{3n} K_1 \left(m_n r\right)\right] \cos\left(m_n z\right), \tag{3.1}$$

$$B_z = \sum_{n=1}^{\infty} m_n \left[a_{3n} I_0 \left(m_n r\right) - b_{3n} K_0 \left(m_n r\right)\right] \sin\left(m_n z\right), \tag{3.2}$$

where r is the radial coordinate and z is the axial coordinate as shown in Fig. 3.1, and

$$m_n = (2n - 1)\pi / t_p. \tag{3.3}$$

$I_0(m_n r), I_1(m_n r)$ are modified Bessel functions of the first kind, and $K_0(m_n r), K_1(m_n r)$ are modified Bessel functions of the second kind, for order zero and one, respectively. a_{3n} and b_{3n} are coefficients of the harmonic, and their values are dependent upon the machine topology, PM material properties and geometric parameters of linear machine. Based on Lorentz law, the axial force exerted on a winding with current density of J, is derived as

$$F_{wp} = -8\pi J \sum_{n=1}^{\infty} K_n \sin\left(m_n \left(z - \frac{\tau_{wp}}{2}\right)\right), \tag{3.4}$$

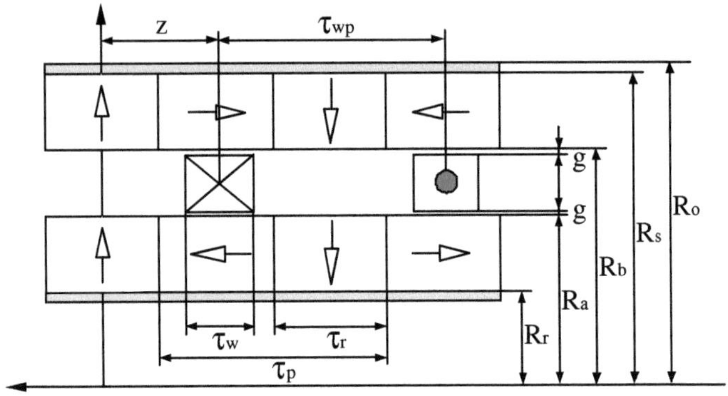

Fig. 3.1 Machine structure with single-phase winding

where

$$K_n = \sin\left(m_n \frac{\tau_w}{2}\right) \sin\left(m_n \frac{\tau_{wp}}{2}\right) K_{rn},$$

$$K_{rn} = \int_{R_i}^{R_b} r\left[a_{3n} I_1\left(m_n r\right) + b_{3n} K_1\left(m_n r\right)\right] dr. \tag{3.5}$$

τ_p is the winding pitch, and τ_{wp} is defined as

$$R_i = R_a + g, \tag{3.6}$$

where R_a is the outer radius of the internal PMs, R_b is the inner radius of the external PMs, g is the size of air gap, and τ_w is the axial width of the phase winding per pole. The winding arrangement for a single-phase slotless tubular linear machine is shown in Fig. 3.1. The current density in the winding is given by

$$J = \sqrt{2} J_{rms} \cos \omega t, \tag{3.7}$$

where J_{rms} is the root mean square (RMS) value and ω is the angular frequency of current inputs. Since the armature moves in synchronism with the AC frequency, we have

$$\omega t = \frac{\pi z}{\tau_p}. \tag{3.8}$$

Thus, from Eq. (3.4), the force generated by a single-phase PM tubular linear machine is

$$F_{wp} = -8\sqrt{2}\pi J_{rms} \sum_{n=1}^{\infty} K_n \sin\left(m_n\left(z - \frac{\tau_{wp}}{2}\right)\right) \cos \omega t. \tag{3.9}$$

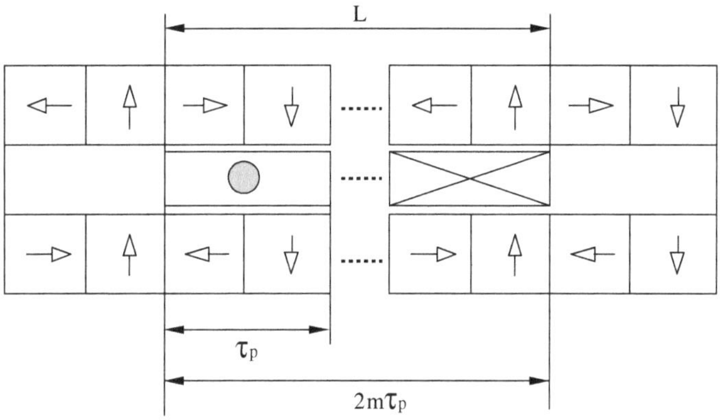

Fig. 3.2 Electromagnetic Linear machine with a single phase current input

As z represents the axial position of the mover, the force output of the machine is related to the motion position. The force formulation can be simplified to

$$F_{wp} = -8\sqrt{2}\pi J_{rms} \sum_{n=1}^{\infty} K_{rn} \cos(m_n z) \cos\left(\frac{\pi z}{\tau_p}\right). \tag{3.10}$$

Full pitch winding is employed to increase force output, which means that the distance between two coils of a phase winding, τ_{wp}, is equal to pole pitch, τ_p. Structure of single-phase machine with multi-pair pole configuration is shown in Fig. 3.2. Therefore, for one-phase current input, axial width of the coil, τ_w, is set equal to the pole pitch, τ_p to take full advantage of magnetic energy and generate a consistent unidirectional force. In this case, the slotless armature consisting of a number of coils with a length of L is inevitably a multiple of τ_p. L can be represented with

$$L = 2p\tau_p, \tag{3.11}$$

where $2p$ is the number of poles. Therefore, for a finite armature with fixed length L, it can mount on either one or more pole pairs respecting to distinctive pole pitch. The force output for a single-phase machine with a coreless armature of L can be derived as

$$F_t = -8p\pi J \sum_{n=1}^{\infty} K_n \sin\left(m_n \left(z - \frac{L}{2p}\right)\right), \tag{3.12}$$

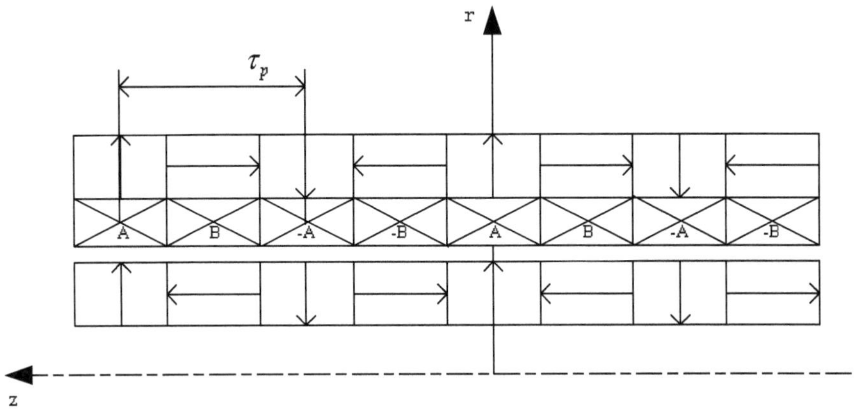

Fig. 3.3 Double-phase winding pattern

where K_n can be simplified to

$$K_n = \sin^2(m_n\tau_p)K_{rn},$$
$$K_{rn} = \int_{R_i}^{R_b} r\left[a_{3n}I_1(m_nr) + b_{3n}K_1(m_nr)\right]dr. \tag{3.13}$$

3.2.3 Thrust for Double-Phase Winding Pattern

The winding pattern for a double-phase slotless tubular linear machine is presented in Fig. 3.3. The winding width is the same for the two phases. The input current densities in phases A and B are

$$J_A = \sqrt{2}J_{rms}\cos wt,$$
$$J_B = \sqrt{2}J_{rms}\sin wt. \tag{3.14}$$

From this pattern, winding pitch, τ_{wp}, equals to pole pitch, τ_p. The thrust generated by each phase is given by

$$F_A = 8\sqrt{2}\pi J_{rms}\left(\sum_{n=1}^{\infty} K_n \sin m_n\left(z - \frac{\tau_p}{2}\right)\cos\frac{\pi z}{\tau_p}\right),$$
$$F_B = 8\sqrt{2}\pi J_{rms}\left(\sum_{n=1}^{\infty} K_n \sin m_n\left(z - \tau_p\right)\sin\frac{\pi z}{\tau_p}\right). \tag{3.15}$$

Therefore, the total force is obtained as

$$F = F_A + F_B = 8\sqrt{2}\pi J_{rms} \sum_{n=1}^{\infty} K_n \left[\sin m_n \left(z - \frac{\tau_p}{2}\right) \cos \frac{\pi z}{\tau_p} \right.$$
$$\left. + \sin m_n \left(z - \tau_p\right) \sin \frac{\pi z}{\tau_p} \right]. \tag{3.16}$$

It can be simplified to

$$F = F_1 + \sum_{n=2}^{\infty} F_n \cos \left[(2n - 1 + (-1)^n) \frac{\pi z}{\tau_p} \right], \tag{3.17}$$

where F_1 is the constant thrust caused by the fundamental term of radial magnetic harmonic components, and F_n is the magnitude of the force ripple created by the $(2n-1)^{\text{th}}$ harmonic component. It is noticed that there is no constant term for the force generated by single-phase winding pattern. F_1 and F_n in Eq. (3.17) are

$$\begin{aligned} F_1 &= -8\sqrt{2}\pi J_{rms} K_1, \\ F_n &= (-1)^n 8\sqrt{2}\pi J_{rms} K_n \quad n = 2, 3, \ldots \end{aligned} \tag{3.18}$$

respectively. The force ripple can be represented as a function of z as

$$F_r = \sum_{n=2}^{\infty} F_n \cos \left[(2n - 1 + (-1)^n) \frac{\pi z}{\tau_p} \right] \quad n = 2, 3, \ldots \tag{3.19}$$

The maximum force ripple produced by the double-phase winding pattern can then be computed with

$$F_{rm} = 8\sqrt{2}\pi J_{rms} \sqrt{\sum_{n=2k}^{\infty} (K_n - K_{n+1})^2}. \tag{3.20}$$

The relative force ripple is defined as the ratio of maximum force ripple and the fundamental term, i.e.,

$$F_{rr} = \frac{F_{rm}}{F_1} = \frac{\sqrt{\sum_{n=2k}^{\infty} (K_n - K_{n+1})^2}}{K_1} \quad k = 1, 2, \ldots \tag{3.21}$$

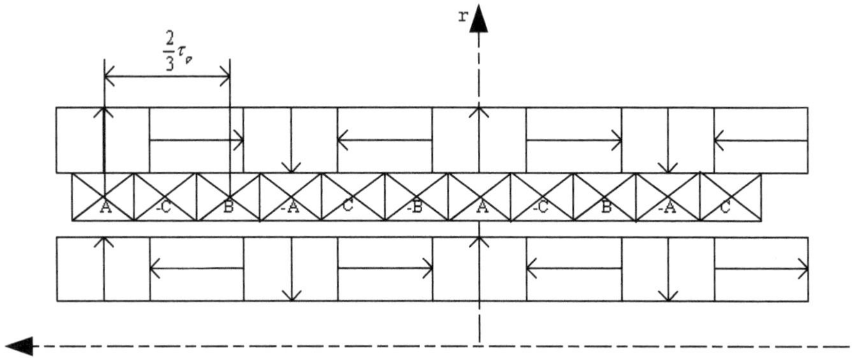

Fig. 3.4 Three phase winding structure

3.2.4 Thrust for Three-Phase Winding Pattern

The three-phase winding pattern of the linear machine is presented in Fig. 3.4. The three phases are separated by two-thirds of the pole pitch in turns. Similar to the double-phase pattern, a three-phase winding comprises a few coils connected in series. In three-phase winding structure, the winding pitch, τ_{wp}, is equal to the pole pitch, τ_p, and each coil occupies one-third of a pole pitch. The current densities in the three phases are given by

$$\begin{aligned} J_A &= \sqrt{2}J_{rms}\cos\left(\omega t + \frac{2}{3}\pi\right), \\ J_B &= \sqrt{2}J_{rms}\cos\omega t, \\ J_C &= \sqrt{2}J_{rms}\cos\left(\omega t - \frac{2}{3}\pi\right). \end{aligned} \tag{3.22}$$

Besides the current inputs, the force output of the linear machine is also related to the starting point of mover. Three typical starting positions of the mover are studied in this chapter, i.e., Position A in Fig. 3.5, Position B in Fig. 3.6, and Position C in Fig. 3.7. For these three positions, the center lines of winding A, B, and C are aligned with the r coordinate. In following subsection, the force output and force ripple for the three position are formulated.

3.2.4.1 Starting Position A

For starting position A in Fig. 3.5, the force generated by the three phases is

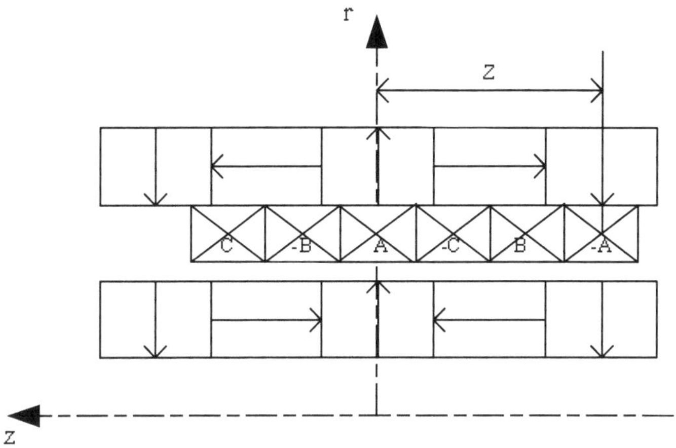

Fig. 3.5 Starting position A of mover

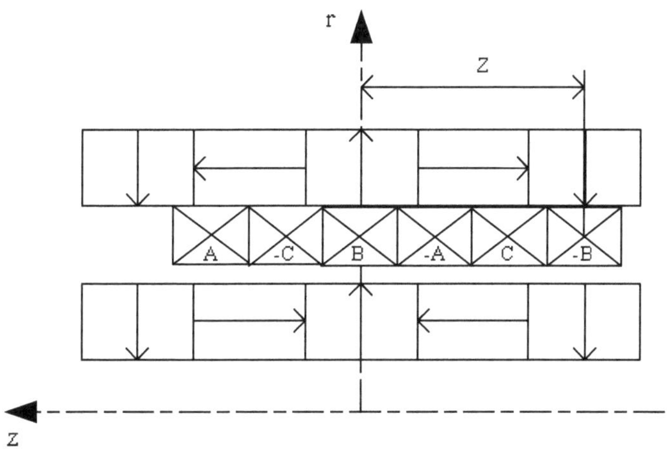

Fig. 3.6 Starting position B of the machine mover

$$
\begin{aligned}
F_A &= 8\sqrt{2}\pi J_{rms} \sum_{n=1}^{\infty} K_n \sin\left(m_n\left(z - \frac{\tau_p}{2}\right)\right) \cos(\omega t + \frac{2}{3}\pi), \\
F_B &= 8\sqrt{2}\pi J_{rms} \sum_{n=1}^{\infty} K_n \sin\left(m_n\left(z + \frac{5\tau_p}{6}\right)\right) \cos \omega t, \\
F_C &= 8\sqrt{2}\pi J_{rms} \sum_{n=1}^{\infty} K_n \sin\left(m_n\left(z + \frac{\tau_p}{6}\right)\right) \cos\left(\omega t - \frac{2}{3}\pi\right).
\end{aligned}
\tag{3.23}
$$

Therefore, the total thrust is

$$\begin{aligned} F &= F_A + F_B + F_C \\ &= 8\sqrt{2}\pi Jrms \sum_{n=1}^{\infty} K_n \Big[\sin\left(m_n\left(z - \frac{\tau_p}{2}\right)\right)\cos\left(\omega t + \frac{2}{3}\pi\right) \\ &+ \sin\left(m_n\left(z + \frac{5\tau_p}{6}\right)\right)\cos\omega t + \sin\left(m_n\left(z + \frac{\tau_p}{6}\right)\right)\cos\left(\omega t - \frac{2}{3}\pi\right)\Big]. \end{aligned} \tag{3.24}$$

It can be further simplified to

$$\begin{aligned} F &= F_1 + \sum_{n=3k}^{\infty} F_{n0} \sin\left[(2n)\left(\frac{\pi z}{\tau_p}\right) + \frac{\pi}{6}\right] \\ &+ \sum_{n=3k+1}^{\infty} F_{n1} \sin\left[(2n-2)\left(\frac{\pi z}{\tau_p}\right) - \frac{\pi}{6}\right] \quad k = 1, 2 \ldots \end{aligned} \tag{3.25}$$

F_1, F_{n0} and F_{n1} are calculated from

$$\begin{aligned} F_1 &= -6\sqrt{2}\pi J_{rms} K_1, \\ F_{n0} &= 12\sqrt{2}\pi J_{rms} K_n (-1)^{k+1}, \\ F_{n1} &= 12\sqrt{2}\pi J_{rms} K_n (-1)^{k+1}. \end{aligned} \tag{3.26}$$

Equation (3.25) indicates that the force ripple generated by the third harmonic term in the radial magnetic field is equal to zero. The maximum force ripple is thus

$$F_{rm} = \sqrt{\sum_{n=3k}^{\infty} \frac{3}{4}(K_n + K_{n+1})^2 + \frac{1}{4}(K_n - K_{n+1})^2} \quad k = 1, 2 \ldots . \tag{3.27}$$

Therefore, the relative force ripple of the tubular linear machine is derived as

$$F_{rr} = \frac{F_{rm}}{F_1} = \frac{\sqrt{\sum_{n=3k}^{\infty} (3/4)(K_n + K_{n+1})^2 + (1/4)(K_n - K_{n+1})^2}}{K_1} \tag{3.28}$$

3.2.4.2 Starting Position B

The starting position B of the mover in the linear machine is illustrated in Fig. 3.6. The force generated by phases of A, B, and C are

$$F_A = 8\sqrt{2}\pi J_{rms}\left(\sum_{n=1}^{\infty} K_n \sin m_n \left(z+\frac{\tau_p}{6}\right)\cos\left(\frac{\pi z}{\tau_p}+\frac{2}{3}\pi\right)\right),$$

$$F_B = 8\sqrt{2}\pi J_{rms}\left(\sum_{n=1}^{\infty} K_n \sin m_n \left(z-\frac{\tau_p}{2}\right)\cos\left(\frac{\pi z}{\tau_p}\right)\right), \tag{3.29}$$

$$F_C = 8\sqrt{2}\pi J_{rms}\left(\sum_{n=1}^{\infty} K_n \sin m_n \left(z+\frac{5\tau_p}{6}\right)\cos\left(\frac{\pi z}{\tau_p}-\frac{2}{3}\pi\right)\right).$$

The total thrust force is obtained as

$$\begin{aligned} F &= F_A + F_B + F_C \\ &= 8\sqrt{2}\pi J_{rms}\sum_{n=1}^{\infty} K_n \Big[\sin m_n \left(z+\frac{\tau_p}{6}\right)\cos\left(\frac{\pi z}{\tau_p}+\frac{2}{3}\pi\right) \\ &\quad +\sin m_n\left(z-\frac{\tau_p}{2}\right)\cos\left(\frac{\pi z}{\tau_p}\right)+\sin m_n\left(z+\frac{5\tau_p}{6}\right)\cos\left(\frac{\pi z}{\tau_p}-\frac{2}{3}\pi\right)\Big]. \end{aligned} \tag{3.30}$$

It can be further simplified to

$$F = F_1 + \sum_{n=3k}^{\infty} F_{n0}\cos\left[(2n)\left(\frac{\pi z}{\tau_p}\right)\right] + \sum_{n=3k+1}^{\infty} F_{n1}\cos\left[(2n-2)\left(\frac{\pi z}{\tau_p}\right)\right] k = 1, 2 \ldots, \tag{3.31}$$

where F_1, F_{n0} and F_{n1} are calculated with

$$\begin{aligned} F_1 &= 12\sqrt{2}\pi JrmsK_1, \\ F_{n0} &= 12\sqrt{2}\pi JrmsK_n(-1)^k, \\ F_{n1} &= 12\sqrt{2}\pi JrmsK_n(-1)^{k+1}. \end{aligned} \tag{3.32}$$

Equation (3.31) shows that the force ripple caused by the third-order harmonics term of the radial magnetic field is equal to zero. The force ripple is derived as

$$F_r = \sum_{n=3k}^{\infty} F_{n0}\cos\left[(2n)\left(\frac{\pi z}{\tau_p}\right)\right] + \sum_{n=3k+1}^{\infty} F_{n1}\cos\left[(2n-2)\left(\frac{\pi z}{\tau_p}\right)\right] \quad k = 1, 2, \ldots. \tag{3.33}$$

The maximum force ripple is

$$F_{rm} = 12\sqrt{2}\pi J_{rms}\sqrt{\sum_{n=3k}^{\infty}(K_n - K_{n+1})^2}. \tag{3.34}$$

And the relative force ripple is

$$F_{rr} = \frac{F_{rm}}{F_1} = \frac{\sqrt{\sum\limits_{n=3k}^{\infty} (K_n - K_{n+1})^2}}{K_1} \quad k = 1, 2 \ldots. \tag{3.35}$$

3.2.4.3 Starting Position C

The starting position C of the mover is shown in Fig. 3.7. The forces generated by the phases of A, B and C are respectively given by

$$\begin{aligned} F_A &= 8\sqrt{2}\pi J_{rms}\left(\sum_{n=1}^{\infty} K_n \sin\left(m_n\left(z + \frac{5\tau_p}{6}\right)\right)\cos\left(\frac{\pi z}{\tau_p} + \frac{2}{3}\pi\right)\right), \\ F_B &= 8\sqrt{2}\pi J_{rms}\left(\sum_{n=1}^{\infty} K_n \sin\left(m_n\left(z + \frac{\tau_p}{6}\right)\right)\cos\left(\frac{\pi z}{\tau_p}\right)\right), \\ F_C &= 8\sqrt{2}\pi J_{rms}\left(\sum_{n=1}^{\infty} K_n \sin\left(m_n\left(z - \frac{\tau_p}{2}\right)\right)\cos\left(\frac{\pi z}{\tau_p} - \frac{2}{3}\pi\right)\right). \end{aligned} \tag{3.36}$$

Correspondingly, the total force is

$$\begin{aligned} F &= F_A + F_B + F_C \\ &= 8\sqrt{2}\pi Jrms \sum_{n=1}^{\infty} K_n[\sin\left(m_n\left(z + \tfrac{5\tau_p}{6}\right)\right)cos\left(\omega t + \tfrac{2}{3}\pi\right)] \\ &\quad + \sin(m_n(z + \tfrac{\tau_p}{6}))\cos\omega t + \sin(m_n(z - \tfrac{\tau_p}{2}))\cos(\omega t - \tfrac{2}{3}\pi)]. \end{aligned}$$

It can be represented with a compact form as

$$\begin{aligned} F = F_1 &+ \sum_{n=3k}^{\infty} F_{n0}\cos\left[(2n)\left(\frac{\pi z}{\tau_p}\right) + \frac{\pi}{3}\right] \\ &+ \sum_{n=3k+1}^{\infty} F_{n1}\sin\left[(2n-2)\left(\frac{\pi z}{\tau_p}\right) + \frac{\pi}{6}\right] \quad k = 1, 2 \ldots. \end{aligned} \tag{3.37}$$

F_1, F_{n0} and F_{n1} are obtained from

$$\begin{aligned} F1 &= 6\sqrt{2}\pi J_{rms}K_1, \\ F_{n0} &= 12\sqrt{2}\pi J_{rms}K_n(-1)^{k+1}, \\ F_{n1} &= 12\sqrt{2}\pi J_{rms}K_n(-1)^{k+1}. \end{aligned} \tag{3.38}$$

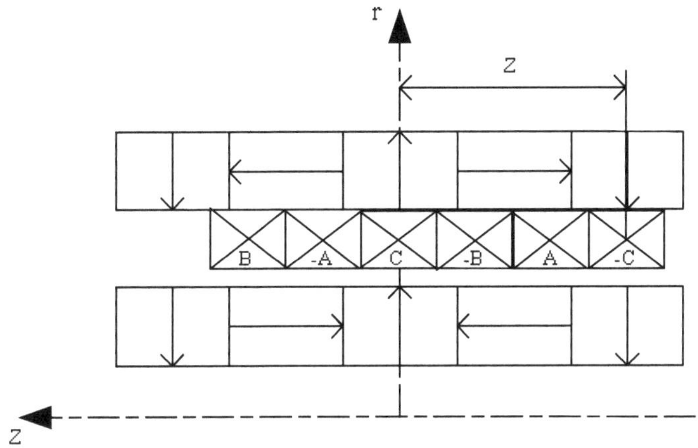

Fig. 3.7 Starting position C of the machine mover

Again, Eq. (3.37) shows that the force ripple by the third harmonic of the radial magnetic field is zero. The total force ripple is

$$FR = \sum_{n=3k}^{\infty} F_{n0} \cos\left[(2n)\left(\frac{\pi z}{\tau_p}\right) + \frac{\pi}{3}\right] + \sum_{n=3k+1}^{\infty} F_{n1} \sin\left[(2n-2)\left(\frac{\pi z}{\tau_p}\right) + \frac{\pi}{6}\right] \quad k = 1, 2, \ldots \tag{3.39}$$

The maximum force ripple is

$$F_{rm} = \sqrt{\sum_{n=3k}^{\infty} \frac{1}{4}(K_n + K_{n+1})^2 + \frac{3}{4}(K_n - K_{n+1})^2} \quad k = 1, 2 \ldots. \tag{3.40}$$

And the relative force ripple is

$$RFR = \frac{MFR}{F_1} = \frac{\sqrt{\sum_{n=3k}^{\infty} \frac{1}{4}(K_n + K_{n+1})^2 + \frac{3}{4}(K_n - K_{n+1})^2}}{K_1}. \tag{3.41}$$

3.3 Simulation and Validation of Force Model

Based on the derived mathematical models, the force outputs for tubular linear machines with single-phase, double-phase, and three-phase winding patterns are simulated. In addition, numerical results are obtained and employed to validate the analytical model of force output, so that it can be used for the performance analysis and structure design purpose.

3.3.1 Force Variation for Single-Phase Machine

The force output with respect to the axial motion for a single-phase linear machine is illustrated in Fig. 3.8. The maximum force is produced when the coil is exactly aligned with the radially magnetized PMs, whereas the minimum one occurs when the coil is aligned with the axially magnetized PMs. The pitch of the magnets is the same as that of coils, which leads to the force variation period consistent to τ_p and the positive peak values in Fig. 3.8. It is found that the constant component in the total force of single-phase machines is equal to zero. In addition, numerical results are computed with Ansoft to compare with the analytical models. It shows that the two sets of data fit with each other closely. Therefore, the derived force model can be employed for parameter analysis and design of the electromagnetic tubular linear machines.

3.3.2 Force Variation for Double-Phase Machine

The force created by double-phase machine is shown in Fig. 3.9. Similarly, the maximum force happens when the coil is exactly aligned with the radially magnetized PMs (Fig. 3.3). It is because the radial component of the magnetic field and the current input arrive the maximum values at this position. The constant component is

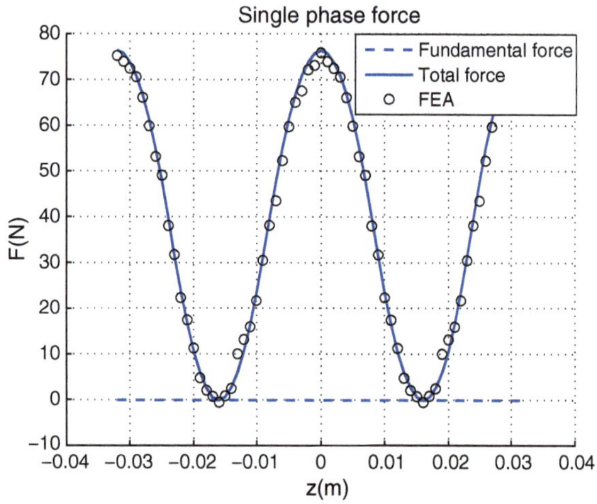

Fig. 3.8 Force output generated by single-phase machine

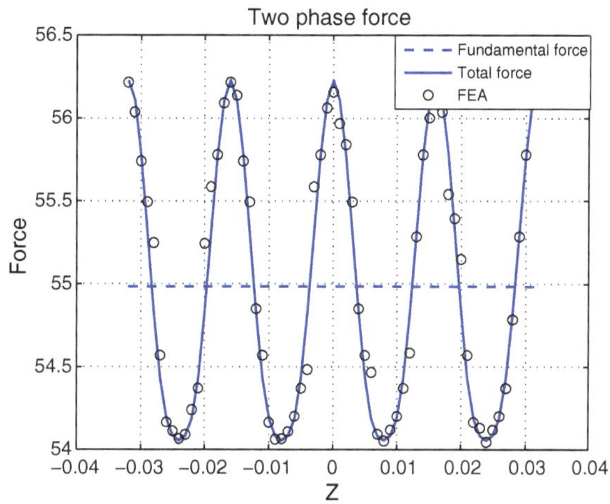

Fig. 3.9 Force output generated by double-phase machine

found in the total force as indicated by the dotted line in the figure. The period of force variation is determined by magnetization and winding patterns. As the distance between two coils is equal to half of the period of magnetization, the force output varies in a period of $\tau_p/2$. Similarly, the numerical results are compared with the analytical model, which indicates that the analytical model fits with the numerical results well and could be employed for parameter design of the linear machines.

3.3.3 Force Variation for Three-Phase Machine

The force variation for three-phase linear machine is also simulated with analytical models, including force output for starting position A in Fig. 3.10, starting position B in Fig. 3.11 and starting position C in Fig. 3.12. It is found that, the force output of a three-phase machine varies in a period of $\tau_p/3$ regardless its starting position. The variation period is determined by the winding pitch. When the mover moves a distance of $\tau_p/3$, the force output created by winding phases changes in turn but the total force is equal to the previous one. For starting position B, the maximum force results from the interaction of the peak current and a maximum magnetic flux density in $\kappa\tau_p$, where κ is an integral number. Specifically, when phase B winding is aligned with the radially magnetized PMs, the force output reaches the maximum value. However, for other two starting positions A and C, the current input and the flux density cannot arrive the maximum values simultaneously wherever the mover is. Therefore, the force output of starting position B is much larger than the rest two cases. Due to the symmetric winding phases, the force variation for A and C are similar with each other.

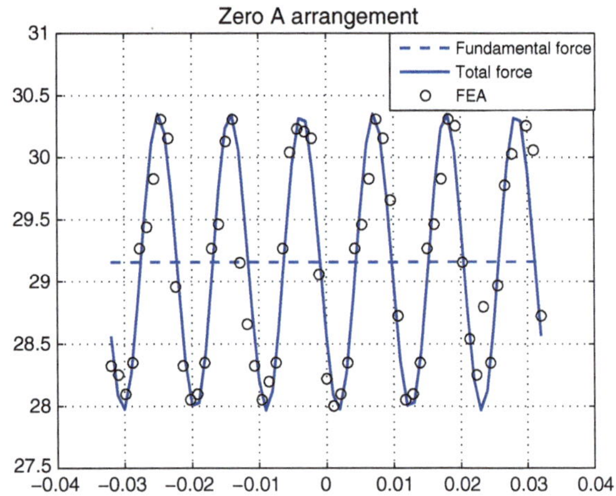

Fig. 3.10 Force output for starting position A

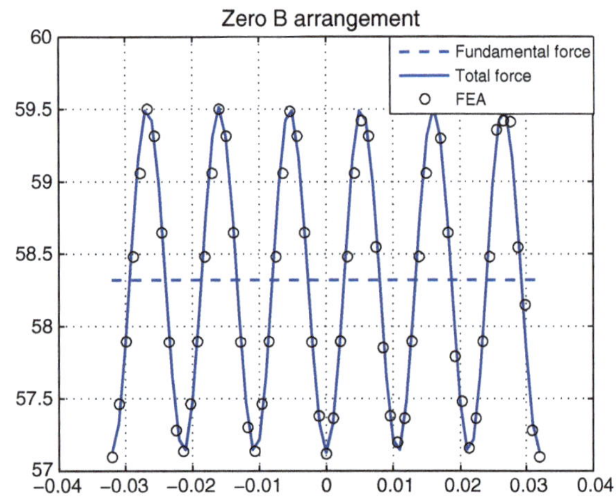

Fig. 3.11 Force output for starting position B

Based on the above simulation results and corresponding analysis, we may have information as follows.

1. For the same electrical inputs and magnetic field, the force output of a three-phase machine is higher than that of the double-phase machine by 6 percent. And the force of a single-phase machine is higher than that of the three-phase on by 12.6 %.
2. For three-phase machines, the starting position B can produce high force output that is almost twice of those of other two positions.
3. For the three-phase machines, the third-order harmonics of the magnetic field does not cause any force ripple. And for single-phase machines, there is no constant component in the force output. Hence, the three-phase machines have lower

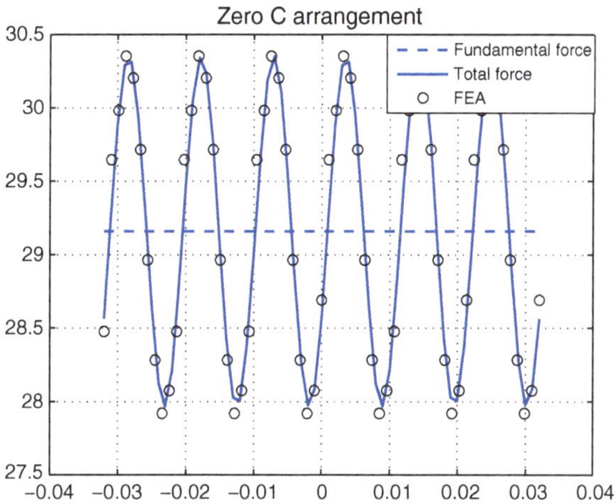

Fig. 3.12 Force output for starting position C

force ripple compared to other two types, which is favorable for good dynamic performance of linear machine.

In general, a three-phase tubular linear machine has a better performance than other two peers considering the force capability and the possible instability caused by force ripple. Furthermore, integrated power modules are widely available for three-phase brushless machine, which greatly benefits the design and commercialization of this type of linear machines. In subsequent discussion, we will mainly focus on the study of three-phase tubular linear machines.

3.4 Conclusion

Based on the flux field models from magnetic vector potential, the force output for linear machines with single-phase, double-phase and three-phase winding patterns is formulated analytically. By using the derived mathematic models, the force output for different winding phases is simulated. Furthermore, numerical results are obtained to validate the derived force models. It shows that the analytical models fit with the numerical results well.

References

1. Yan L, Chen IM, Lim CK et al (2011) Design, modeling and experiments of 3-DOF electromagnetic spherical actuators. Springer, Berlin
2. Gieras JF, Wing M et al (1998) Permanent magnet motor technology. Marcel Dekker Inc, New York

3. Im DH, Kim CE (1994) Finite element force calculation of a linear induction motor taking account of the movement. IEEE Trans Magn 30(5):3495–3498
4. Adamiak K, Mizia J, Dawson GE et al (1987) Finite element force calculation in linear induction machines. IEEE Trans Magn 23(5):3005–3007
5. Zhou G, Huang X, Jiang H et al (2011) Analysis method to a Halbach PM ironless linear motor with trapezoid windings. IEEE Trans Magn 47(10):4167–4170
6. Krause PC, Wasynczuk O et al (1989) Electromechanical Motion Devices. McGraw-Hill Inc, New York USA
7. Cho Y, Lee J, Koo D (1995) The analysis of the static thrust of a hybrid type double-sided linear pulse motor. IEEE Trans Magn 31(3):2084–2087
8. Zare-Bazghaleh A, Meshkatoddini MR, Fallah-Choolabi E (2016) Force study of single-sided linear induction motor. IEEE Trans Plasma Sci 44(5):849–856
9. Sadiku NOM (2001) Elements of Electromagnetics. Oxford University Press, UK 2001
10. Wang J, Wang W, Jewell GW et al (1998) Design and experimental characterisation of a linear reciprocating generator. IEEE Proc-Electr Power Appl 145(6):509–518
11. Wang J, Jewell GW, Howe D (1999) A general framework for the analysis and design of tubular linear permanent magnet machines. IEEE Trans Magn 35(3):1986–2000
12. Wang J, Jewell GW, Howe D (2001) Design optimisation and comparison of tubular permanent magnet machine topologies. IEEE Proc Electr Power Appl 148(5):456–464
13. Wang J, Howe D, Jewell GW (2002) An improved axially magnetised tubular permanent magnet machine topology. In: Proceedings of IET internatinal conference on power electronics, machines and drives, pp. 303–308, Bath, UK, 4–7 Sept 2002
14. Chen X, Zhu ZQ, Howe D (2009) Modeling and analysis of a tubular oscillating permanent-magnet actuator. IEEE Trans Ind Appl 45(6):1961–1970
15. Chen X, Zhu ZQ (2011) Analytical determination of optimal split ratio of E-core permanent magnet linear oscillating actuators. IEEE Trans Ind Appl 47(1):25–33
16. Boules N (1984) Two-dimensional field analysis of cylindrical machines with permanent magnet excitation. IEEE Trans Ind Appl IA–20(5):1267–1277

Chapter 4
Armature Reaction Field and Inductance

Abstract Analysis of armature reaction field and inductance is extremely important for design and control implementation of electromagnetic machines. So far, most studies have focused on magnetic field generated by permanent magnet poles, whereas less work has been done on armature reaction field. This chapter proposes a novel analytical modeling method to predict the armature reaction field of a coreless permanent magnet tubular linear machine (PMTLM) with dual Halbach array. Unlike conventional modeling approach, the proposed method formulates the armature reaction field for electromagnetic machines with finite length, so that the analytical modeling precision can be improved. In addition, winding inductance is also analytically formulated to facilitate dynamic motion control based on the reaction field solutions. Numerical result is subsequently obtained with finite element method, and employed to validate the derived analytical models.

4.1 Introduction

Armature is an important component that influences the system output significantly. When the machine is energized, the current in windings shifts and distorts the main flux, which is known as armature reaction [1]. The magnetic field on load is therefore a synthesis of PM field and armature reaction field. Although armature reaction field is relatively low, it may cause instability and tracking error, particularly for high-precision system [2–4]. Therefore, it is important to analyze the armature reaction field precisely and, thus, to improve the control performance of permanent magnet linear machines. Analytical modeling based on harmonic expansion is generally an efficient method to study armature reaction [5]. It is more precise in the description of field distribution than conventional permeance models [6–8] and current sheet models [9–11]. It is also time saving and effective in design optimization of electromagnetic machines compared with finite element method [12–14]. Researchers have obtained analytical models of armature reaction fields on the premise of infinite length of electric motors. For example, Wang et al. presented accurate analytical expression of armature reaction field for a tubular PMLM with infinite length in [15]. Amara et al. derived the armature reaction field of a tubular PMLM by assuming

L. Yan et al., *Electromagnetic Linear Machines with Dual Halbach Array*, DOI 10.1007/978-981-10-2309-5_4

infinite length of an electric linear machine [16, 17]. The assumption of infinite length of linear motor apparently simplifies the formulation procedure. However, it may also reduce the modeling precision and unavoidably influences subsequent system design and motion control implementation of electric machines. Therefore, this chapter proposes an analytical modeling method to describe magnetic field distribution of armature reaction in moving-coil PMTLMs by taking finite motor length into consideration. This method absorbs the merits of analytical model based on harmonic expansion and permeance model. The established model of armature reaction field is validated by FEM solutions, and utilized to formulate winding inductance. It can benefit dynamic analysis and control of electric linear machines [16, 18]. A research prototype with dual Halbach array and single phase current is developed, and experiments are conducted on both armature reaction field and inductance to verify the developed analytical models. Along with the open-circuit analytical model, the proposed armature reaction field model can be employed for prediction of on load field distribution of linear machines [19].

4.2 Formulation of Armature Reaction Field

4.2.1 Assumptions

Following assumptions are made to facilitate the subsequent formulation of armature reaction field.

- The magnetic permeability of back iron is infinitely large with respect to that of free space.
- Laminated irons are used, and thus eddy currents in both stator and mover of the linear machines are ignorable.
- The relative permeability of rare earth permanent magnet is approximately equal to that of free space.
- All materials are isotropic in terms of magnetic property.

4.2.2 Governing Equations

Based on Maxwell equations and Coulomb gauge, the governing equation is derived as

$$\nabla^2 \mathbf{A} = -\mu_0 \mu_r \mathbf{J}. \tag{4.1}$$

The machine space under study is separated into two types. Type 1 is the central part without hatched lines in Fig. 4.1 with the mover included, and type 2 is the part filled with horizontal hatched lines. Because only magnetic flux in the winding region

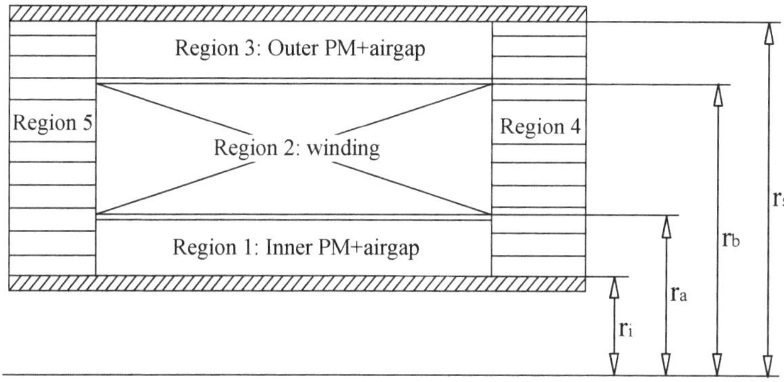

Fig. 4.1 Five-region division of the machine space

affects the output performance of the linear machine, this section mainly focuses on the study of central part. The hatched part will be employed for the study of inductance computation in subsequent section. The central part can be divided into two groups according to the current values. Specifically, Group 1 includes two layers of PMs and air gaps (Region 1 and 3), and Group 2 is the current region (Region 2) with a single phase current input. The two groups are governed by Laplace's and Poisson's equations as follows.

$$\begin{aligned} &\nabla^2 \mathbf{A}_i = 0, \quad \mathrm{i} = \{1, 3\} \\ &\nabla^2 \mathbf{A}_2 = -\mu_0 \mathbf{J}. \end{aligned} \tag{4.2}$$

Due to the axially symmetric structure of the PMTLM, the magnetic vector potential **A** only has θ component. Therefore, in the cylindrical coordinate system CS1 shown in Fig. 4.2, $\mathbf{A} = (0, 0, A_\theta)$. Since the machine is symmetric about z-axis, A_θ is only a function of r and z. The governing equation in the cylindrical coordinate system CS1 is

$$\begin{aligned} &\frac{\partial^2 A_{i\theta}}{\partial z^2} + \frac{\partial}{\partial r}\left(\frac{1}{r}\frac{\partial}{\partial r}(r A_{i\theta})\right) = 0, \\ &\frac{\partial^2 A_{2\theta}}{\partial z^2} + \frac{\partial}{\partial r}\left(\frac{1}{r}\frac{\partial}{\partial r}(r A_{2\theta})\right) = -\mu_0 J. \end{aligned} \tag{4.3}$$

4.2.3 Current Source System

The coreless PMTLM adopts single phase winding for short stroke, ease of control, and low cost [20]. The same approach could be extended to electromagnetic machines with multiple phase windings. The current density distribution, accompanying with the winding configuration, is shown in Fig. 4.3. The harmonic expansion of the current density J is

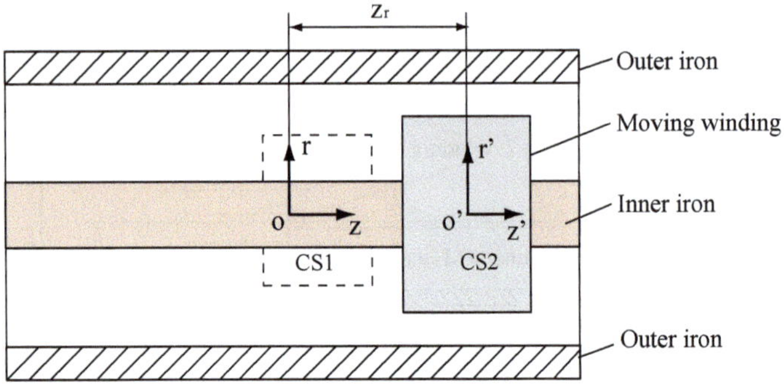

Fig. 4.2 Cylindrical coordinate systems CS1 and CS2

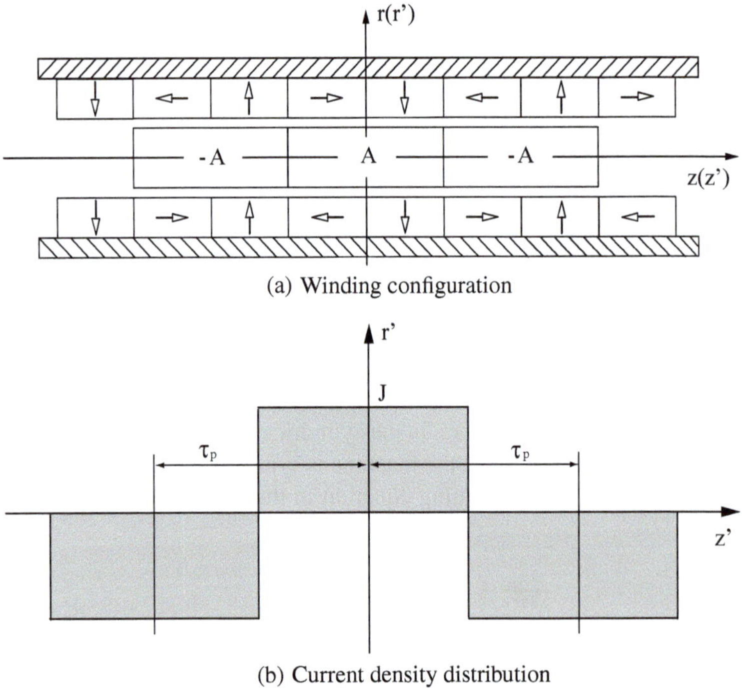

Fig. 4.3 Winding configuration and current density distribution

$$
\begin{aligned}
J_s(z') &= \sum_{n=1}^{\infty} \frac{4J}{(2n-1)\pi} \sin\left(\frac{(2n-1)\pi}{2}\right) \cos\left(\frac{(2n-1)\pi}{\tau_p} z'\right) \\
&= \sum_{n=1}^{\infty} J_n \cos\left(\frac{(2n-1)\pi}{\tau_p} z'\right),
\end{aligned}
\tag{4.4}
$$

where z' represents the axial position in the coordinate system CS2, as shown in Fig. 4.2. Its origin is at the geometry center of the moving winding. The transformation between CS2 and the static coordinate system CS1 fixed at the motor center is

$$r' = r, \quad z' = z + z_r, \tag{4.5}$$

where z_r represents the mover position in CS1.

4.2.4 Analytical Solutions of Magnetic Field

4.2.4.1 Magnetic Field of Noncurrent Region

Magnetic field of no current region is the solution to the Laplace's equation in Eq. (4.3). By taking variable separation, we assume that $A_{i\theta}(r, z) = R(r)Z(z)$. Laplace's equation is simplified to

$$\frac{1}{z(z)}\frac{\partial Z^2(z)}{\partial z^2} = k^2,$$
$$r^2\frac{\partial R^2(r)}{\partial r^2} + r\frac{\partial R(r)}{\partial r} + (k^2r^2 - 1)R(r) = 0,$$

where k^2 is an eigenvalue. The general solution is

$$A_{i\theta} = \left(p_i r + q_i \frac{1}{r}\right)(s_i + t_i z) + \sum_{n=1}^{\infty} \{[a_{in} J_1(k_n r) + b_{in} Y_1(k_n r)]$$
$$[c_{in} e^{k_n z} + d_{in} e^{-k_n z}] + [e_{in} I_1(m_n r) + f_{in} K_1(m_n r)]$$
$$[g_{in} \cos(m_n z) + h_{in} \sin(m_n z)]\}, \tag{4.6}$$

where $J_1(kr)$ and $Y_1(kr)$ are Bessel functions of the first kind and second kind, respectively, and $I_1(mr)$ and $K_1(mr)$ are modified Bessel functions of the first kind and second kind, respectively. As the axial flux density is periodic, the exponential terms are

$$c_{in} = 0, \quad d_{in} = 0.$$

By taking the curl of magnetic vector potential **A**, the radial and axial flux densities are obtained as

$$B_{ir} = -\left(p_i r + q_i \frac{1}{r}\right) t_i + \sum_{n=1}^{\infty} m_n [e_{in} I_1(m_n r) + f_{in} K_1(m_n r)]$$
$$\sin(m_n z'),$$
$$B_{iz} = 2(s_i + t_i) p_i + \sum_{n=1}^{\infty} m_n [e_{in} I_0(m_n r) - f_{in} K_0(m_n r)]$$
$$\cos(m_n z'). \tag{4.7}$$

4.2.4.2 Magnetic Field of Current Region

Magnetic field distribution in the current region is the solution to the Poisson's equation of Eq. (4.3). The solution to Poisson's equation is the superposition of the homogeneous solution to Laplace's equation and a particular solution to the Poisson's equation. Substituting Eq. (4.4) into the right side of Poisson's equation, and using variable separation again yield

$$
\begin{gathered}
Z(z) = \cos(m_n z'),\\
(m_n r)^2 \frac{\partial^2 R}{\partial (m_n r)^2} + (m_n r)\frac{\partial R}{\partial (m_n r)} - [1 + (m_n r)^2] = P_n \frac{2(m_n r)^2}{\pi},
\end{gathered}
$$

where m_n and P_n are calculated with

$$
m_n = \tfrac{(2n-1)\pi}{\tau_p}, \quad P_n = -\tfrac{\pi \mu_0 J_n}{2{m_n}^2}.
$$

A particular solution to Poisson's equation by employing the Struve function [21] is given as

$$
A_{2\theta}{}^* = \sum_{n=1}^{\infty} L_1(m_n r) P_n \cos(m_n z'). \tag{4.8}
$$

Therefore, the magnetic flux density in the current region is

$$
\begin{aligned}
B_{2r} &= -\left(p_2 r + q_2 \tfrac{1}{r}\right) t_2 + \textstyle\sum_{n=1}^{\infty} m_n [e_{2n} I_1(m_n r)\\
&\quad + f_{2n} K_1(m_n r) + L_1(m_n r) P_n] \sin(m_n z'),\\
B_{2z} &= 2(s_2 + t_2) p_2 + \textstyle\sum_{n=1}^{\infty} m_n [e_{2n} I_0(m_n r)\\
&\quad - f_{2n} K_0(m_n r) + L_0(m_n r) P_n] \cos(m_n z').
\end{aligned} \tag{4.9}
$$

4.2.4.3 Boundary Conditions

Boundary conditions (BCs) are required to determine the unique solution of the magnetic field. Subjecting to Gauss's law and Ampere's law, magnetic field obeys certain rules along the boundary of two different materials. The component of magnetic flux density perpendicular to the boundary is always continuous, i.e., $B_{1n} = B_{2n}$. The tangential component of magnetic field is discontinuous by the amount of surface current at the boundary, i.e., $H_{1t} - H_{2t} = K$. In particular, in this chapter, boundary conditions of electromagnetic field are

$$
\begin{aligned}
&B_{1z}|_{r=r_i} = 0, && B_{3z}|_{r=r_o} = 0,\\
&B_{1r}|_{r=r_a} = B_{2r}|_{r=r_a}, && H_{1z}|_{r=r_a} = H_{2z}|_{r=r_a},\\
&B_{2r}|_{r=r_b} = B_{3r}|_{r=r_b}, && H_{2z}|_{r=r_b} = H_{3z}|_{r=r_b}.
\end{aligned} \tag{4.10}
$$

Through solving the boundary condition equations, the following constraints are obtained

$$\begin{aligned} p_1 = p_2 = p_3 = 0, \\ q_1 t_1 &= q_2 t_2 = q_3 t_3 = -Q, \\ UX = Y, \end{aligned} \tag{4.11}$$

where

$$U = \begin{bmatrix} I_0(r_1) & -K_0(r_1) & 0 & 0 & 0 & 0 \\ I_1(r_2) & K_1(r_2) & -I_1(r_2) & K_1(r_2) & 0 & 0 \\ I_0(r_2) & -K_0(r_2) & -I_0(r_2) & K_0(r_a) & 0 & 0 \\ 0 & 0 & I_1(r_3) & K_1(r_3) & -I_1(r_3) & -K_1(r_3) \\ 0 & 0 & I_0(r_3) & -K_0(r_3) & -I_0(r_3) & K_0(r_3) \\ 0 & 0 & 0 & 0 & I_0(r_4) & -K_0(r_4) \end{bmatrix},$$

$$X = \begin{bmatrix} a_{1n} & b_{1n} & a_{2n} & b_{2n} & a_{3n} & b_{3n} \end{bmatrix}^{\mathrm{T}},$$

$$Y = \begin{bmatrix} 0 & P_n L_1(l_a) & P_n L_0(l_a) & -P_n L_1(l_b) & -P_n L_0(l_b) & 0 \end{bmatrix}^{\mathrm{T}},$$

and in coefficient matrix U,

$$r_1 = m_n r_i, \quad r_2 = m_n r_a, \quad r_3 = m_n r_b, \quad r_4 = m_n r_o.$$

Therefore, the analytical expressions of flux density in regions i ($i = 1, 3$) and 2 can be simplified to

$$\begin{aligned} B_{ir} &= \frac{Q}{r} + \sum_{n=1}^{\infty} m_n [e_{in} I_1(m_n r) + f_{in} K_1(m_n r)] \sin(m_n z'), \\ B_{iz} &= \sum_{n=1}^{\infty} m_n [e_{in} I_0(m_n r) - f_{in} K_0(m_n r)] \cos(m_n z'), \end{aligned} \tag{4.12}$$

and

$$\begin{aligned} B_{2r} &= \frac{Q}{r} + \sum_{n=1}^{\infty} m_n [e_{2n} I_1(m_n r) + f_{2n} K_1(m_n r) \\ &\quad + L_1(m_n r) P_n] \sin(m_n z'), \\ B_{2z} &= \sum_{n=1}^{\infty} m_n [e_{2n} I_0(m_n r) - f_{2n} K_0(m_n r) \\ &\quad + L_0(m_n r) P_n] \cos(m_n z'). \end{aligned} \tag{4.13}$$

To get the value of Q without complex computation, we utilize equivalent magnetic circuit (EMC), instead of analytical modeling based on harmonic expansion, to achieve a close solution. Since the waveform of the flux density is already determined in the analytical expression, Q is tolerable to estimate, and it will not decrease much the precision of the analytical model much.

4.2.4.4 Coefficients Determination

The coil numbers, magnetic flow distribution, and EMC are shown in Fig. 4.4, respectively. For a machine with single phase winding consisting of $(2k+1)$ coils, the magnetomotive force (MMF) generated by the nth coil is

$$F_{\text{n}} = (-1)^n NI, n = -k, -(k-1), \ldots, k-1, k. \tag{4.14}$$

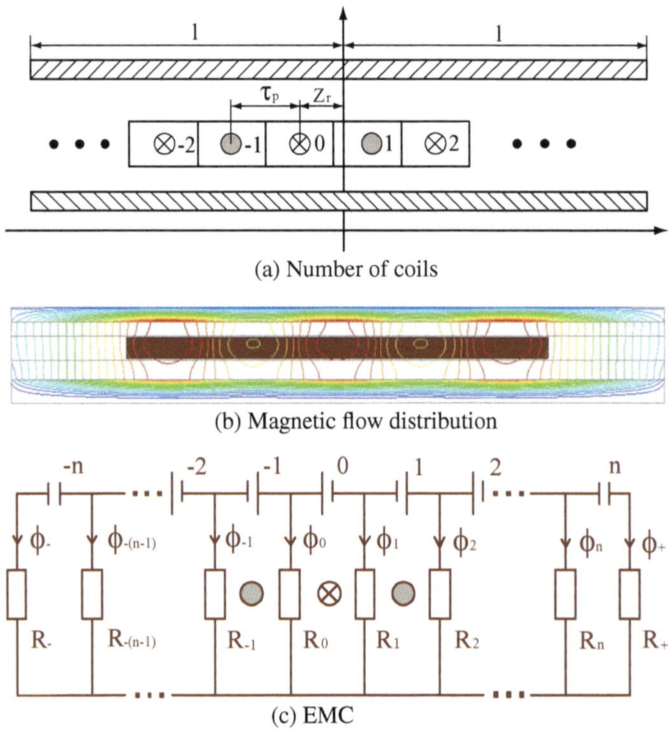

Fig. 4.4 Modeling process of EMC

Magnetic reluctance in the circuit is classified into three types, i.e.,

$$\begin{aligned} R_n &= \frac{g}{2\pi \mu_0 \bar{r} \tau_p}, \\ R_+ &= \frac{g}{2\pi \mu_0 \bar{r}(l - z_r - k\tau_p)}, \\ R_- &= \frac{g}{2\pi \mu_0 \bar{r}(l + z_r - k\tau_p)}, \end{aligned} \tag{4.15}$$

where $\bar{r}$ is the equivalent radii for gap reluctance. It is defined as

$$\bar{r} = \frac{r_o - r_i}{\ln(r_o/r_i)}.$$

The magnetic circuit equations are given as

$$\begin{aligned} \phi_- R_- - \phi_{-(k-1)} R_{-(k-1)} &= F_{-k}, \\ \phi_n R_n - \phi_{n+1} R_{n+1} &= F_n, \\ \phi_k R_k - \phi_+ R_+ &= F_k, \\ \phi_- + \phi_+ + \sum_{n=-(k-1)}^{k} \phi_n &= 0. \end{aligned} \tag{4.16}$$

By solving the preceding equations, the main magnetic flux between coil 1 and coil 0 (ϕ_1), and that between coil 0 and coil -1 (ϕ_0) is related with

$$\frac{\phi_1}{\phi_0} = -\frac{l + (-1)^k z_r}{l - (-1)^k z_r}. \tag{4.17}$$

Meanwhile, ϕ_0 and ϕ_1 can be deduced from the analytical expression of flux density, i.e., (4.9), as

$$\begin{aligned} \phi_1 &= \int_{z'=\tau_p/4}^{z'=3\tau_p/4} 2\pi r B_{2r} dz', \\ \phi_0 &= \int_{z'=-3\tau_p/4}^{z'=-\tau_p/4} 2\pi r B_{2r} dz', \end{aligned} \tag{4.18}$$

where r is an arbitrary number from r_a to r_b. The range of z' in Eq. (4.18) is obtained from the magnetic flow distribution in Fig. 4.4b. By substituting Eq. (4.18) into (4.17), Q is obtained as

$$Q = \frac{2(-1)^k \bar{r} \sum_{n=1}^{\infty} y_n}{l\tau_p} z_r,$$

where

$$y_n = e_{2n} I_1(m_n \bar{r}) + f_{2n} K_1(m_n \bar{r}) + L_1(m_n \bar{r}) P_n.$$

Therefore, all coefficients in analytical models are determined.

4.3 Winding Inductance

Winding inductance influences system dynamics of electromagnetic machines significantly. Because of the employment of single phase winding, only self-inductance exists. Magnetic energy utilized to analyze the winding inductance is

$$W = \iiint\limits_V \frac{\mathbf{B} \bullet \mathbf{H}}{2} dv = \frac{1}{2\mu_0} \iiint\limits_V \mathbf{B}^2 dv. \tag{4.19}$$

For Region 1, 2, and 3 (type 1), flux densities have been formulated, and the sum of energies in these areas is

$$\begin{aligned} W_a &= W_1 + W_2 + W_3 \\ &= \frac{\pi}{\mu_0}[\int_{z_b}^{z_t}\int_{r_i}^{r_a} ({B_{1r}}^2 + {B_{1z}}^2) r' dr' dz' + \int_{z_b}^{z_t}\int_{r_a}^{r_b} ({B_{2r}}^2 + {B_{2z}}^2) \\ & r' dr' dz' + \int_{z_b}^{z_t}\int_{r_b}^{r_o} ({B_{3r}}^2 + {B_{3z}}^2) r' dr' dz'], \end{aligned} \tag{4.20}$$

where the bound for z' is

$$\begin{aligned} z_b &= -(k\tau_p + \tau_p/2), \\ z_t &= k\tau_p + \tau_p/2. \end{aligned} \tag{4.21}$$

For Region 4 and 5 (type 2), the EMC method can provide a field model accurate enough for inductance computation. The energy in type 2 is

$$\begin{aligned} W_b &= \frac{\pi}{\mu_0}[\int_{z_t}^{l-z_r}\int_{r_i}^{r_o} {B_+}^2 r' dr' dz' + \int_{-l-z_r}^{z_b}\int_{r_i}^{r_o} {B_-}^2 r' dr' dz'] \\ &= \frac{({R_o}^2 - {R_i}^2)}{8\pi\mu_0 \bar{r}^2}[\frac{{\phi_+}^2}{l - z_r - z_t} + \frac{{\phi_-}^2}{l + z_r + z_b}]. \end{aligned} \tag{4.22}$$

The magnetic energy resulting from the inductance matrix of the five-coil winding is

$$W = \frac{1}{2}[L]^T[N_k I][L], \tag{4.23}$$

where

$$[L] = \begin{bmatrix} L_{11} & M_{12} & M_{13} & M_{14} & M_{15} \\ M_{21} & L_{22} & M_{23} & M_{24} & M_{25} \\ M_{31} & M_{32} & L_{33} & M_{34} & M_{35} \\ M_{41} & M_{42} & M_{43} & L_{44} & M_{45} \\ M_{51} & M_{52} & M_{53} & M_{54} & L_{55} \end{bmatrix},$$

and

$$[N_k I] = [\, N_k I_1 \; N_k I_2 \; N_k I_3 \; N_k I_4 \; N_k I_5 \,]^T.$$

Assume that the five coils in the winding are completely equivalent in the inductive characteristics, i.e.,

$$\begin{aligned} L_{ii} &= L_{jj}, \quad i, j = 1, 2, \ldots 5, \\ M_{ij} &= M_{ji} = L_{ii}, \quad j \neq i, \end{aligned} \tag{4.24}$$

The self-inductance of a single coil is given. When referring to the principle that the winding inductance is proportional to the square of turns, the self-inductance of the single phase winding with five coils is

$$L_w = 5^2 L_{ii}. \tag{4.25}$$

4.4 Validation by Finite Element Method

Although FEM is time consuming, it can achieve high precision results, taking non-linear characteristics and flux leakage into consideration. Therefore, FEM is utilized to validate the established analytical models, including magnetic field model and inductance model. The structure parameters of a five-coil PMTLM in FEM computation is given in Table 4.1.

Table 4.1 Structure parameters in FEM computation

r_i (mm)	5	r_a (mm)	9	r_b (mm)	14	r_o (mm)	17
τ_p (mm)	18	l (mm)	100	N_k	200	J (A/m^2)	$4e+6$

Fig. 4.5 Armature reaction field variation versus z' validated by FEM

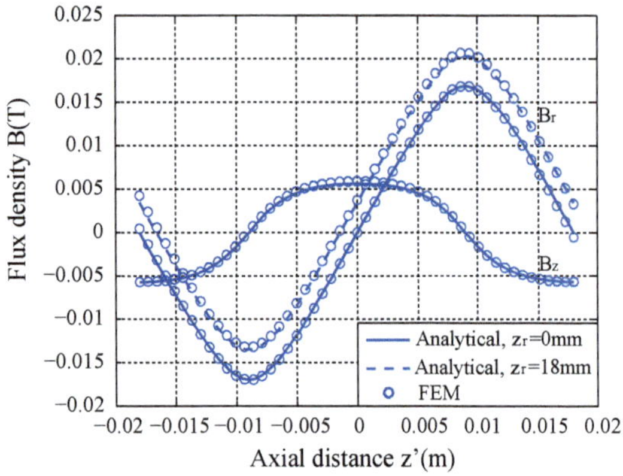

(a) Flux density at $r = (r_i + r_a)/2$

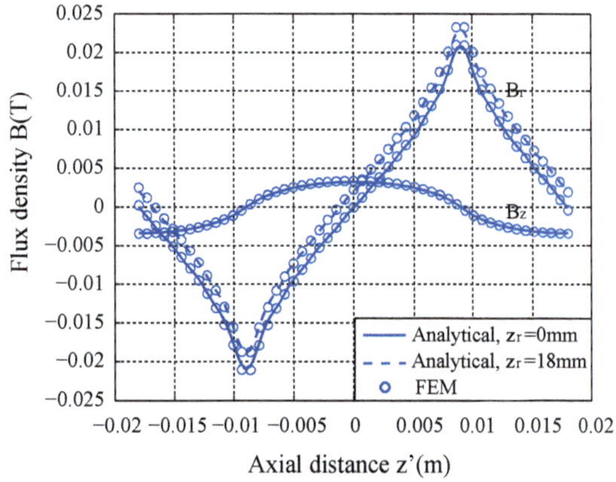

(b) Flux density at $r = (r_a + r_b)/2$

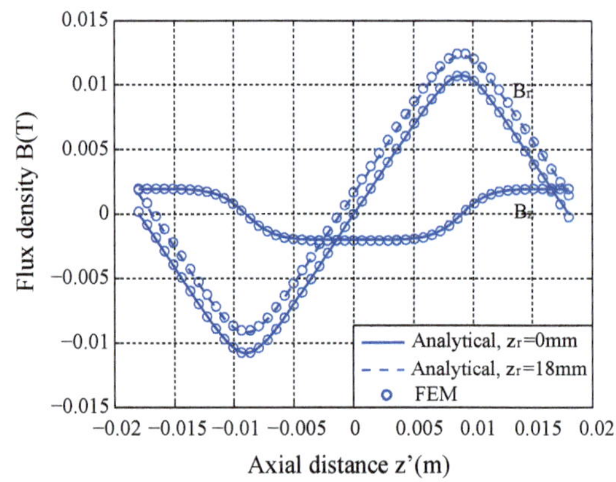

(c) Flux density at $r = (r_b + r_o)/2$

4.4.1 *Armature Reaction Field Validation*

4.4.1.1 Flux Density Versus z'

Figure 4.5 presents the flux density variation, including radial and axial components, versus axial distance z' at $r = (r_i + r_a)/2$, $r = (r_a + r_b)/2$, and $r = (r_b + r_s)/2$, respectively. Because the armature reaction field is influenced by the mover position z_r, the flux density variation for different z_r is also shown in the figure. The solid curve is obtained at $z_r = 0$. It indicates that the winding locates at the center of the machine. The dotted curve is given at $z_r = 18$ mm, a pole pitch distance. It can be found that the mover position z_r only affects the distribution of the radial flux component. It is mainly because magnetic reluctance change caused by mover translation is hardly related to the axial component as indicated in Fig. 4.4. The analytical solutions fit well with the FEM results.

4.4.1.2 Flux Density Versus *r*

Figure 4.6 presents the magnetic flux density variation versus radial distance at $z' = \tau_p/4$ for different mover position $z_r = 0$ and $z_r = \tau_p$. It also shows that the mover translation hardly affects the axial component of the magnetic flux density. The radial component decreases for large value versus radial distance r because of the increased cross section for constant magnetic flux. The analytical solutions are consistent with the FEM results, which indicates that the analytical field model is of high precision comparable with numerical computation.

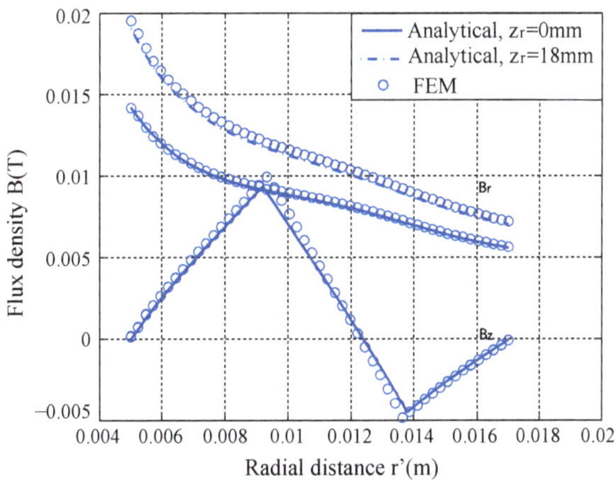

Fig. 4.6 Armature reaction field variation versus r' validated by FEM

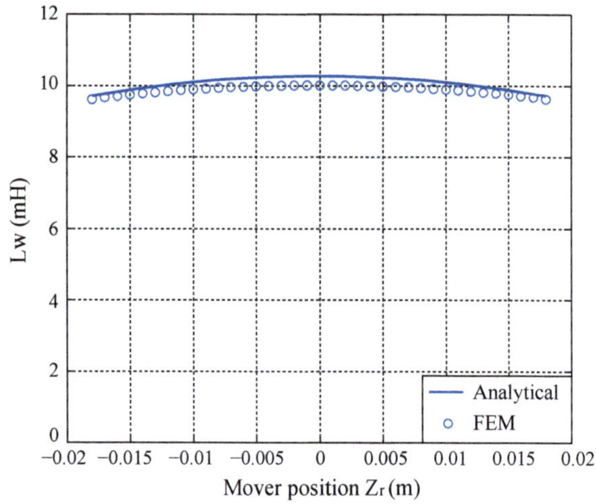

Fig. 4.7 Armature reaction field variation versus r' validated by FEM

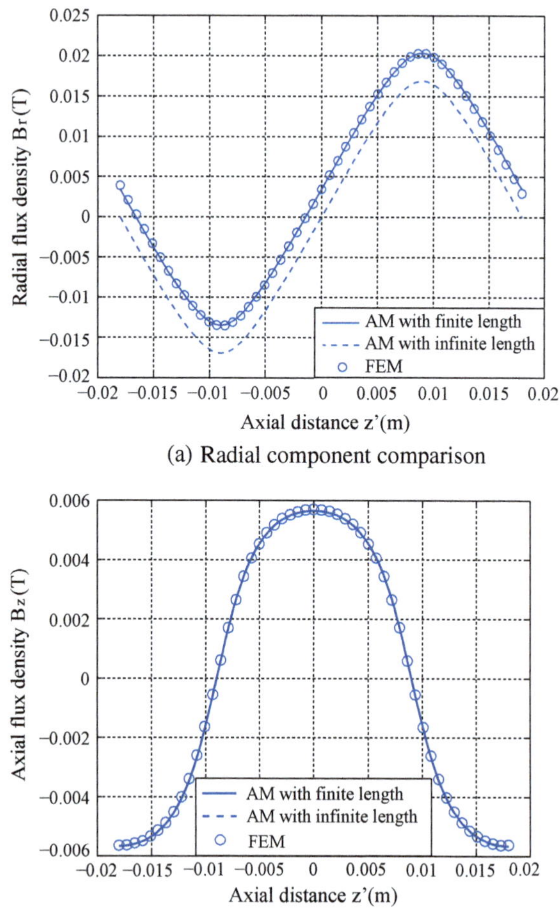

Fig. 4.8 Comparison of magnetic field components of analytical models with FE results at r = $(r_i + r_a)/2$

4.4.2 Inductance Validation

The relationship of self-inductance L_w versus mover position z_r is shown in Fig. 4.7. It is shown that the difference between the analytical solutions and numerical results is very small, which implies that the inductance model can be employed for system dynamic analysis and control implementation. It also, in turn, verifies the high precision of the flux field model. The difference is mainly caused by the EMC method in the computation of magnetic energy computation.

4.4.3 Comparison with Finite Length Model

Magnetic flux density components of armature reaction obtained by the developed mathematical model are compared with conventional analytical model for infinite machine length. The result is presented in Fig. 4.8. It shows that the developed armature reaction field model fits with the numerical computation well, which, in turn, indicates that the proposed method helps to improve mathematical modeling of magnetic field distribution.

4.5 Conclusion

The analysis of armature reaction field is significant for high-precision motion control of electromagnetic machines. To improve the modeling precision, this chapter has proposed one novel method to formulate the armature reaction field of a coreless moving-coil PMTLM by taking finite motor length into consideration. Winding inductance is subsequently analytical models. The comparison shows that the mathematical models fit with numerical results closely. The proposed modeling method is effective and could be implemented into analysis of other electromagnetic machines.

References

1. Jang SM, Jeong SS (2001) Armature reaction effect and inductance of moving coil linear oscillatory actuator with unbalanced magnetic circuit. IEEE Trans Magn 37(4):2847–2850
2. Rigatos GG (2009) Adaptive fuzzy control of DC motors using state and output feedback. Electr Power Syst Res 79(11):1579–1592
3. Shen JX, Tseng KJ (2003) Analyses and compensation of rotor position detection error in sensorless PM brushless DC motor drives. IEEE Trans Energy Convers 18(1):87–93
4. Horng JH (1999) Neural adaptive tracking control of a DC motor. Inf Sci 118(1):1–13
5. Praveen RP, Ravichandran MH, Achari VTS et al (2012) A novel slotless Halbach-array permanent-magnet brushless dc motor for spacecraft applications. IEEE Trans Industr Electron 59(9):3553–3560

6. Vansompel H, Sergeant P, Dupr L (2012) A multilayer 2-DC2-D coupled model for eddy current calculation in the rotor of an axial-flux PM machine. IEEE Trans Energy Convers 27(3):784–791
7. Fujimoto Y, Kominami T, Hamada H (2009) Development and analysis of a high thrust force direct-drive linear actuator. IEEE Trans Industr Electron 56(5):1383–1392
8. Asama J, Hamasaki Y, Oiwa T et al (2013) Proposal and analysis of a novel single-drive bearingless motor. IEEE Trans Industr Electron 60(1):129–138
9. He JL, Levi E, Zabar Z et al (1991) Analysis of induction-type coilgun performance based on cylindrical current sheet model. IEEE Trans Magn 27(1):579–584
10. Zhu ZQ, Howe D, Bolte E et al (1993) Instantaneous magnetic field distribution in brushless permanent magnet DC motors, part I: Open-circuit field. IEEE Trans Magn 29(1):124–135
11. Chan TF, Lai LL, Xie S (2009) Field computation for an axial flux permanent-magnet synchronous generator. IEEE Trans Energy Convers 24(1):1–11
12. Bianchi N, Bolognani S, Corte DD et al (2003) Tubular linear permanent magnet motors: an overall comparison. IEEE Trans Ind Appl 39(2):466–475
13. Lou Z, Yu K, Wang L et al (2010) Two-reaction theory of homopolar inductor alternator. IEEE Trans Energy Convers 25(3):677–679
14. Torkaman H, Afjei E, Toulabi MS (2012) New double-layer-per-phase isolated switched reluctance motor: concept, numerical analysis, and experimental confirmation. IEEE Trans Industr Electron 59(2):830–838
15. Wang J, Jewell GW, Howe D (1999) A general framework for the analysis and design of tubular linear permanent magnet machines. IEEE Trans Magn 35(3):1986–2000
16. Amara Y, Barakat G, Reghem P (2011) Armature reaction magnetic field of tubular linear surface-inset permanent-magnet machines. IEEE Trans Magn 47(4):805–811
17. Amara Y, Barakat G (2010) Analytical modeling of magnetic field in surface mounted permanent-magnet tubular linear machines. IEEE Trans Magn 46(11):3870–3884
18. Cupertino F, Giangrande P, Pellegrino G et al (2011) End effects in linear tubular motors and compensated position sensorless control based on pulsating voltage injection. IEEE Trans Industr Electron 58(2):494–502
19. Bellara A, Tiegna H, Amara Y et al (2012) On load analytical modelling of the magnetic field for axial flux surface-inset permanent magnet machines with semi-closed slots. In: Electrical machines (ICEM), Marseille, France, pp 2852–2858. 2–5 September 2012
20. Chen YT, Chiu CL, Jhang YR et al (2013) A driver for the single-phase brushless DC fan motor with hybrid winding structure. IEEE Trans Industr Electron 60(10):4369–4375
21. Struve functions (2014). http://www.maplesoft.com/support/help/Maple/view.aspx?path=StruveH

Chapter 5
Machine Design and Optimization

Abstract This chapter studies the parameter optimization and back iron influence of tubular linear machines with dual Halbach array. The force output of a single phase permanent magnet tubular linear machine is obtained analytically in Chap. 3. Based on the analytical force model, penalty method and scanning method are employed to conduct parameter optimization to maximize the force output for a given machine volume. Then the influence of structure parameters on force output and force ripple of a three-phase linear machine has been investigated to select optimized structure parameters. Ferromagnetic materials affect magnetic field distribution and force output of electromagnetic machines significantly. Numerical method in Ansoft environment is employed to compute the magnetic field and force output of linear machines with various types of back iron patterns including internal back iron, external back iron, double-sided back iron, and without back iron.

5.1 Introduction

Of the various linear machine structures, tubular linear machine offers high force density and low detent force because of nonexistence of transverse end effect. However, since a linear machine is directly connected to actuators without torque/force-increasing mechanisms, such as gear reducer, its force density is generally under satisfactory. In order to improve the force density of linear machines, a number of methods have been proposed. Based on the analytical model, optimization methods are required to obtain optimal results. The optimization design of electric machine is generally a complicated problem due to a large number of design parameters and the various relationships among parameters [1]. Consequently, the optimization method must be reliable and universally controllable during solution process. In Sect. 5.2, penalty method is independent of the objective function and offers greater reliability than the more computationally efficient methods which use gradient information [2]. Scanning method is a simple and reliable technique provided that the increment is refining. Section 5.2 carries out both penalty method and scanning method on a single phase PMTLM with dual Halbach array, for the sake of global optimal results. Based on the force models, the influence of structure parameters on force output

L. Yan et al., *Electromagnetic Linear Machines with Dual Halbach Array*, DOI 10.1007/978-981-10-2309-5_5

and force ripple of a three-phase linear machine has been investigated in Sect. 5.3. The study provides an effective tool for comprehensively understanding the operating principle and output performance of linear machines with dual Halbach array. Magnet array affects magnetic field distribution and force output of electromagnetic motors severely. Cylindrical linear Halbach arrays are widely used in tubular linear machines to generate a sinusoidal field and offer high efficiency [3–5]. A slotless tubular linear motor with special PMs is introduced in [6]. It is more effective for enhancing the thrust density and reducing the thrust ripple. The armature structure influences force density significantly. Force density of a slotted iron core-type permanent magnet linear synchronous motor is twice as large as that of coreless type PMLSM [7]. Parameter optimization for the purpose of intensified magnetic field of electromagnetic motors has been done in [8, 9]. In these studies, back irons are assumed to be infinitely permeable, and thus they can affect magnetic path greatly without magnetic saturation. Back iron effect on the magnetic field of a tubular linear machine with radially magnetized permanent magnets is discussed in [10]. In Sect. 5.4, back iron effect on both magnetic field and force output of a tubular linear machine with dual Halbach array is analyzed with numerical calculation. The result helps magnetic circuit optimization and improves working efficiency of electromagnetic linear machines.

5.2 Parameter Design of a Single Phase Machine

The design optimization is implemented on a single phase PMTLM with given specifications listed in Table 5.1. Therefore, from Eq. 3.12, the force output is apparently a function of multivariate including pole pair number p, width of the radial magnets τ_r, inner radius of the external Halbach R_b, outer and inner radius of the internal Halbach R_r and R_a. To make dimensionless of design parameters and simplify the design optimization, five substitutes are employed, i.e., p, $\alpha = \tau_r/\tau_p$, $k_{bs} = R_b/R_s$, $k_{ab} = R_a/R_b$, and $k_{ra} = R_r/R_a$. Therefore, the optimal design can be conducted by considering force output as the objective function and five dimensionless parameters as the design variables.

Table 5.1 Major specifications of a PMTML for optimization

Max radius R_0	18 (mm)
Motor length L_m	72 (mm)
Armature length L	36 (mm)
Air gap g	0.2 (mm)
Thick of iron cover ($R_0 - R_s$)	1 (mm)

5.2.1 Penalty Method

Penalty method is a certain class of algorithms for solving constrained optimization problems. It can switch a constrained optimization problem to a unconstrained problem by adding a term to the objective function. The added term is defined zero when constraints are satisfied and nonzero when the constraints are violated. Therefore, the solution to the new unconstrained problem ideally converges to that of the original constrained problem. The main advantage of this technique is that it is independent of the nature of the objective function and can overcome problems due to discontinuities in the objective function. Therefore, it offers greater reliability than the more computationally efficient methods which use gradient information [11]. As the waveform of force output is hardly dependent of design parameters and it reaches peak at $z = 0$, the optimizing goal can be maximize force output at $z = 0$. The design problem can be replaced with the following constrained problem as

$$\max(F_t(z = 0)), \tag{5.1}$$

subject to

$$\begin{aligned} &\mathrm{p} \in \mathrm{int}eger, \\ &0 < \alpha < 1, \\ &0 < k_{bs} < 1, \\ &0 < k_{ab} < 1, \\ &0 < k_{ra} < 1. \end{aligned} \tag{5.2}$$

As the pole pair number p is integer, the penalty method is repeatedly implemented according to different ps to avoid complex integer penalty method. The objective function is used to identify the optimum design. It is applied to all candidate designs, aiming at locating the design for which the function is minimum. Therefore, for particular p, the objective function of penalty method is

$$f_o = \frac{1}{F_t(z = 0)} + r \sum_{i=1}^{4} \left(\frac{1}{\mathrm{x(i)}} + \frac{1}{1 - x(i)} \right), \tag{5.3}$$

where r is the penalty factor, $x(1) = \alpha, x(2) = k_{bs}, x(3) = k_{ab}, x(4) = k_{ra}$. To eliminate the possibility of a local minimum solution, multiple initial iteration values are preset in the program. The flowchart of penalty method in MATLAB environment is shown in Fig. 5.1. The optimization results from the penalty method for different, ps, are shown in Table 5.2. The pole pair number, p, is chosen less than 9 because too many poles will increase leakage flux and thus decrease the system efficiency. In addition, too many poles will increase manufacturing difficulties and costs. Meanwhile, Table 5.2 has shown the tendency of a reverse relationship between force output and pole numbers. The optimal design is marked with deep color.

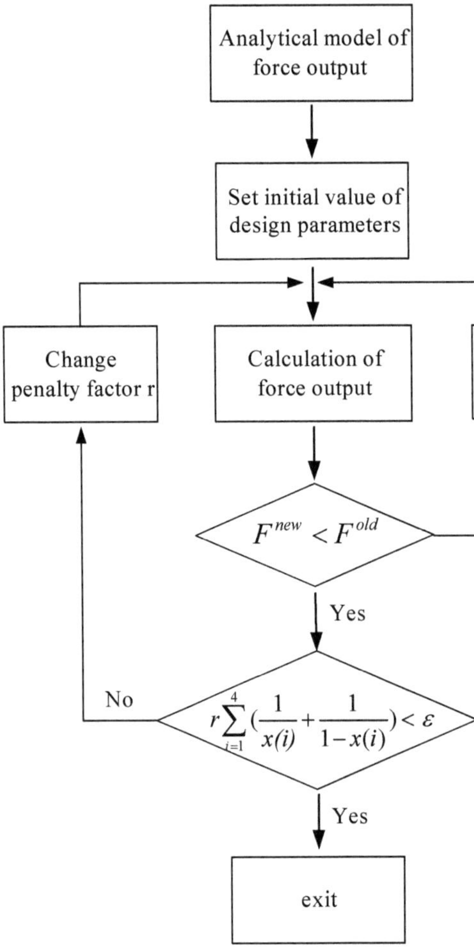

Fig. 5.1 Flowchart of penalty method

Table 5.2 Optimal results from penalty method

m	α	k_{bs}	k_{ab}	k_{ra}	$(F_t(z=0))(N)$
1	0.856	0.849	0.604	0.483	13.681
2	0.741	0.851	0.650	0.368	12.069
3	0.647	0.855	0.698	0.158	10.42
4	0.598	0.863	0.737	0.145	8.321
5	0.545	0.872	0.768	0.227	7.193
6	0.539	0.882	0.794	0.365	6.762
7	0.533	0.891	0.815	0.369	5.239
8	0.528	0.898	0.832	0.436	5.140

5.2.2 Scanning Method

Scanning method is to identify the leading design parameters and sweep them between preset limits, and feasible designs being processed for many possible combinations of the parameters [12]. This approach is easy to implement, requiring only a transfer of all candidate design parameters to the design kernel, which in turn returns an optimal design. The principal disadvantage of the scanning method is the possibility of a combinatorial explosion, especially when the number of candidate designs is enormous. This will lead to huge time-consuming and probably exhaustion of computer memory. For example, if the number of scanning parameters is r and the number of increments is n, then the total number of candidate designs is n^r. Increasing the number of scanning parameters by one, however, leads to the number of designs being multiplied n. Therefore, in order to balance computing time and optimization accuracy, we carry out the procedure repeatedly, starting with broader ranges and larger increments, and eventually converging to a likely optimal result

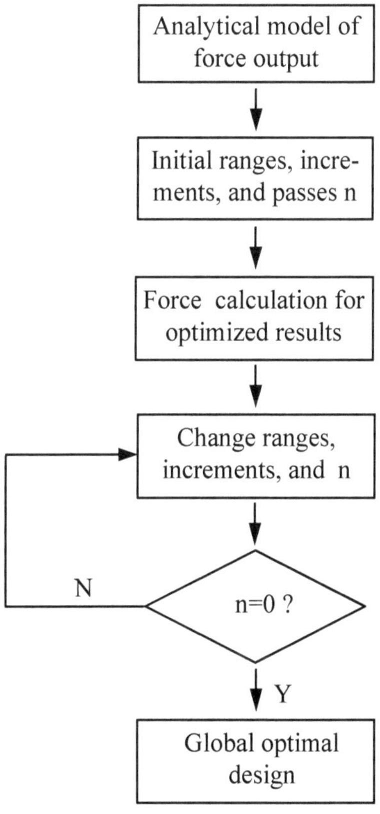

Fig. 5.2 Flowchart of scanning method

Table 5.3 Optimal results from scanning method

m	α	k_{bs}	k_{ab}	k_{ra}	$(F_t(z=0))(N)$
1	0.856	0.848	0.602	0.482	13.6805

Table 5.4 Optimum structure parameters

Max radius R_o	18 mm
Armature length L	36 mm
Pole pairs m	1
Air gap g	0.2 mm
Outer radius of external Halbach R_s	17 mm
Inner radius of external Halbach R_b	14.4 mm
Outer radius of internal Halbach R_a	8.7 mm
Inner radius of internal Halbach R_r	4.2 mm
Axial width of radial PM τ_r	15.4 mm

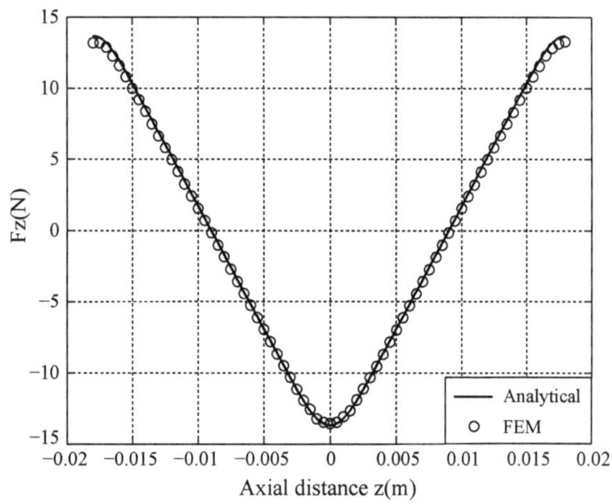

Fig. 5.3 Force output validation of an optimum motor

with successive design passes. By considering that force output as the objective function is a continuous function, the technique is reasonable and reliable. The flowchart of the scanning method is shown in Fig. 5.2. This section executes three times of scanning process with respective increment of 0.05, 0.01, and 0.001. The optimal results are listed in Table 5.3.

5.2.3 Results and Discussion

Optimal parameters and corresponding forces from penalty method and scanning method are consistent, which means the results are reliable. Structure parameters of the optimized machine are shown in Table 5.4. Force output of a optimum design based on analytical model is validated by carrying out an FEM computation. Figure 5.3 shows that the analytical results are very close to the actual response, which in turn verifies the accuracy of the analytical model, and thus the optimal results.

5.3 Parameter Design of Three-Phase Machine

By using the validated analytical models of total force and force ripple, the parameter design of three-phase tubular linear machine can be carried out. In this section, the parameter design targets to maximize the force output and reduce the force ripple to less than 5 % in a limited machine volume.

5.3.1 *Geometric Parameters of the Linear Machine*

Equations 3.30 and 3.35 can be used to predict the force output and force ripple of the PM linear machine. The major geometric parameters of the dual Halbach linear machine are illustrated in Fig. 5.4. R_r and R_a are inner radius and outer radius of internal PMs, respectively. R_b and R_s are the inner radius and outer radius of external PMs, respectively. τ_r is the width of radially magnetized PMs, while τ_z is the width of axially magnetized PMs. As sum of τ_r and τ_z is equal to the pole pitch, τ_p, only τ_r and τ_p will be discussed here. For slotless machines, we assume that the influence of the end effects associated with the finite armature length is negligible. The influence of air gap is not as significant as that of slotted machines [13–15]. It can be considered as a constant. Normally, the maximum machine size is limited by particular applications. In this study, the max radius of the linear machine, R_s, is fixed at 40 mm.

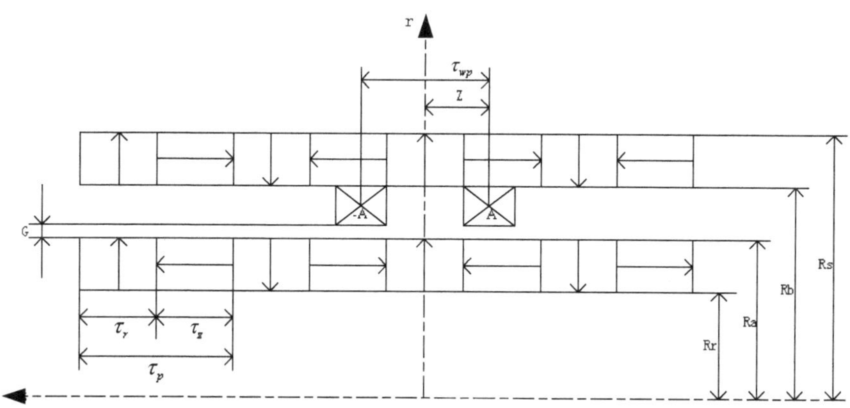

Fig. 5.4 Design parameters of machine equipped with dual Halbach arrays

The force output and force ripple can be seen as functions of the rest five geometric parameters, R_b, R_a, R_r, τ_p and τ_r. The technology of nondimensionalization is utilized in this study, i.e., the five-dimensional parameters are replaced with nondimensionalization parameters, $k_{bs} = R_b/R_s$, $k_{ab} = R_a/R_b$, $k_{ra} = R_r/R_a$, $k_{ps} = \tau_p/R_s$, and $k_{rp} = \tau_r/\tau_p$. Nondimensionalization is the partial or full removal of units from a mathematical equation by a suitable substitution of variables. It is generally used in mechanical, economic, or sociological areas to evaluate the weight of different inputs, such as the oil gap in the hydrodynamic lubrication system. By using the technique of nondimensionalization, the behavior of the system can be analyzed regardless of the units used to measure the variables.

5.3.2 *Interdependence of Structure Parameter*

For structure design of the linear machine, we must consider the interdependence of all five parameters. In other words, the influence between the dimensionless parameters, k_{bs}, k_{ab}, k_{ra}, k_{ps}, and k_{rp} needs to be analyzed. In this study, if one parameter has little influence on the rest ones, it is denoted as independent parameter. Otherwise, it is denoted dependant parameter. The dependence of the geometric parameters is discussed as follows.

5.3.2.1 Relationship Between k_{bs} and Other Parameters

Figure 5.5 shows how the fundamental force output varies with respect to k_{bs}, for different values of k_{ab}, k_{ra}, k_{ps}, and k_{rp}. The fundamental force is defined as the time integral of force output divided by the complete period, i.e.,

$$\widetilde{F} = \int_{-\tau_p}^{\tau_p} F dz / 2\tau_p. \tag{5.4}$$

From Figs. 5.5b, d, it is found that regardless the ratio of k_{ra} and k_{rp}, the position for the maximum value of fundamental force does not change much, which indicates that these two parameters have no significant impact on k_{bs}. However, Fig. 5.5a, c show that for different values of k_{ab} or k_{ps}, the peak value of average force is achieved at different ratios of k_{bs}. Therefore, k_{ab} and k_{ps} have close relationship with k_{bs}.

5.3.2.2 Relationship Between k_{ab} and Other Parameters

Figure 5.6 presents how the force varies as a function of k_{ab} for different values of k_{ra}, k_{ps}, k_{rp}, respectively. From these three figures, it is found that the ratios of k_{ab}

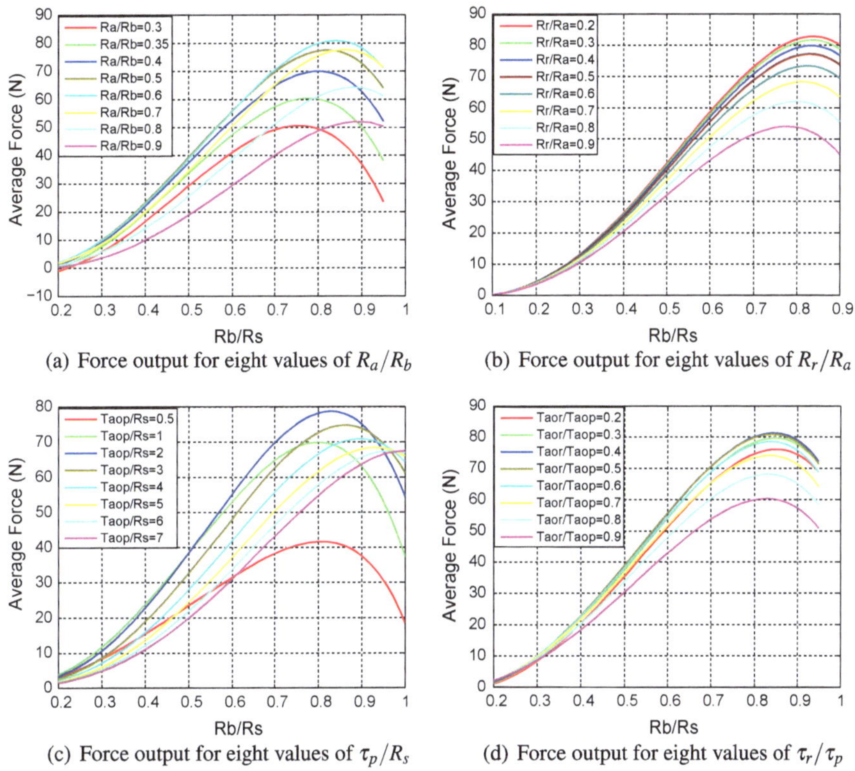

(a) Force output for eight values of R_a/R_b

(b) Force output for eight values of R_r/R_a

(c) Force output for eight values of τ_p/R_s

(d) Force output for eight values of τ_r/τ_p

Fig. 5.5 Force output as a function of R_b/R_s

for the maximum average force does not change much with respect to k_{ra}, k_{ps} and k_{rp}, which indicates that the three parameters have no close interdependence on k_{ab}.

5.3.2.3 Relationship Between k_{ra} and Other Parameters

The force variation with respect to k_{ra} for different values of k_{ps}, k_{rp} is presented in Fig. 5.7a, b, respectively. Similarly, from Figs. 5.5b, 5.6a and 5.7a, b it is found that the ratio k_{ra} is not closely related to other four dimensionless parameters.

5.3.2.4 Relationship Between k_{ps} and Other Parameters

Figure 5.8 illustrates the force variation with respect to k_{ps} for values of k_{rp}. It indicates that for different values of k_{rp}, the peak force is achieved at different ratios of k_{ps}. Therefore, k_{ps} is closely dependent on k_{rp}. Figure 5.5c presents the dependence of

Fig. 5.6 Force output as a function of R_a/R_b

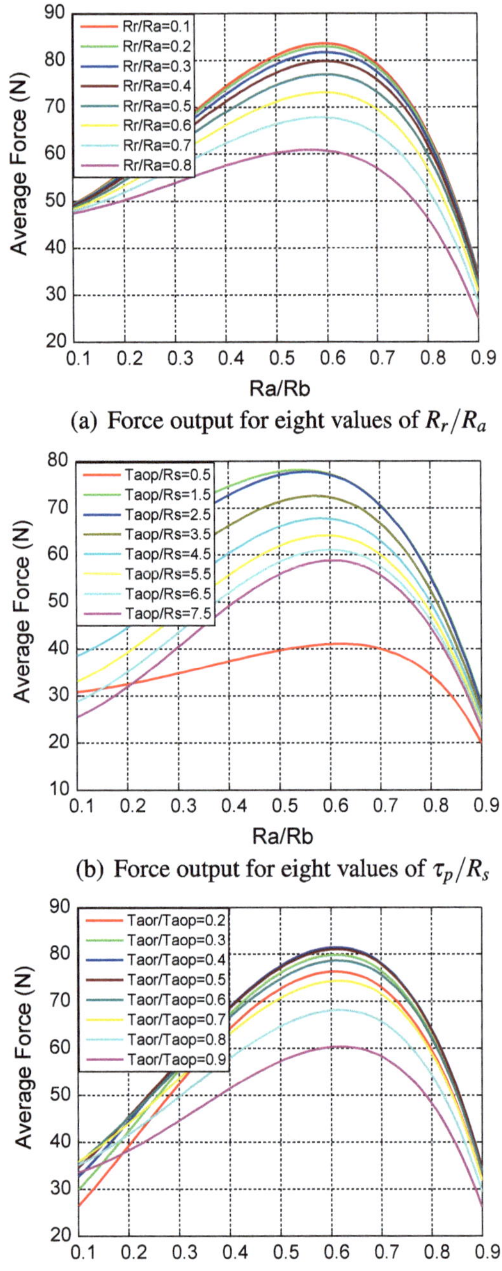

(a) Force output for eight values of R_r/R_a

(b) Force output for eight values of τ_p/R_s

(c) Force output for eight values of τ_r/τ_p

Fig. 5.7 Force output as a function of R_r/R_a

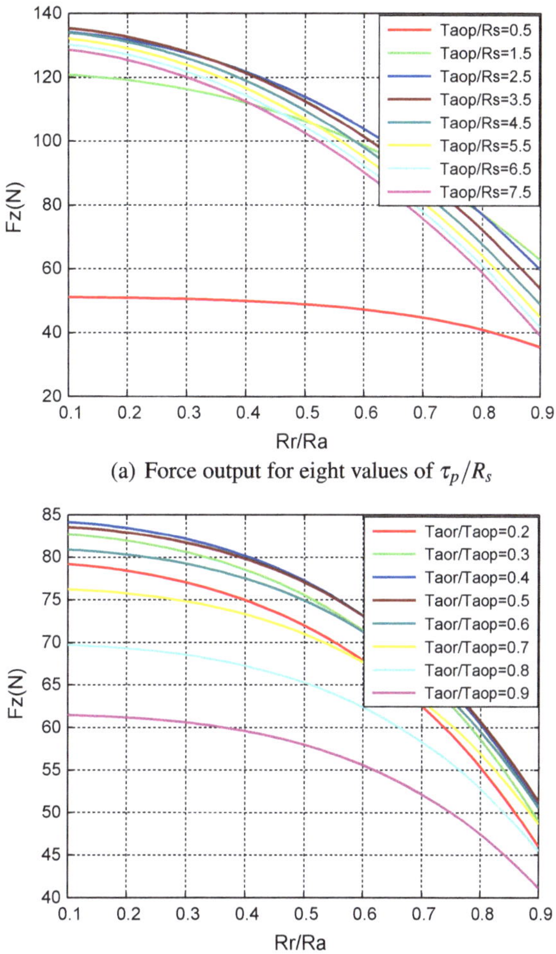

(a) Force output for eight values of τ_p/R_s

(b) Force output for eight values of τ_r/τ_p

k_{ps} and k_{bs}, while Figs. 5.6b and 5.7a have shown the independence of k_{ps}, k_{ab}, and k_{ra}. Therefore, the value of k_{ps} has close relationship with k_{bs} and k_{rp}.

5.3.2.5 Relationship Summary of All Parameters

By summarizing the above analysis, the close relationship exists between parameter pairs of k_{ab} and k_{bs}, k_{ps} and k_{bs}, k_{rp} and k_{ps}. And the rest parameter pairs do not have close relationship, i.e., k_{ra} and k_{bs}, k_{ra} and k_{ab}, k_{ps} and k_{ab}, k_{rp} and k_{ab}, k_{ps} and k_{ra}, k_{rp} and k_{ra}.

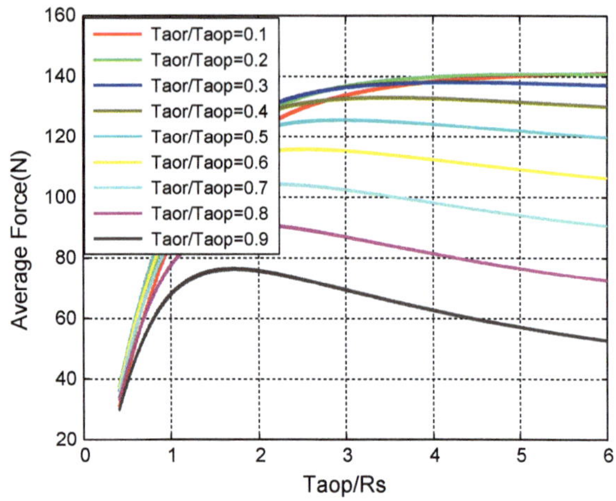

Fig. 5.8 Force output as a function of τ_p/R_s

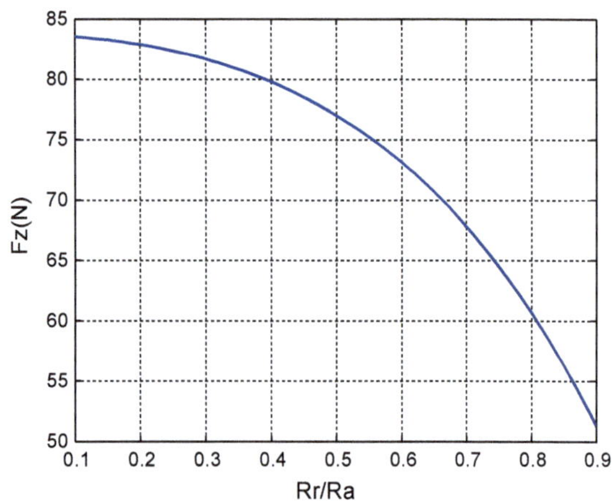

(a) Force output as function of R_r/R_a

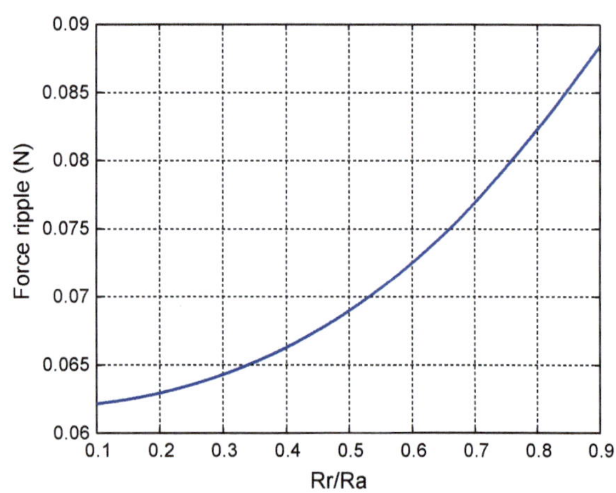

(b) Force ripple as function of R_r/R_a

Fig. 5.9 Force output and force ripple as functions of k_{ra}

Fig. 5.10 Force output and force ripple versus k_{ps} and k_{rp}, at $k_{bs} = 0.85$, $k_{ab} = 0.6$, $k_{ra} = 0.35$

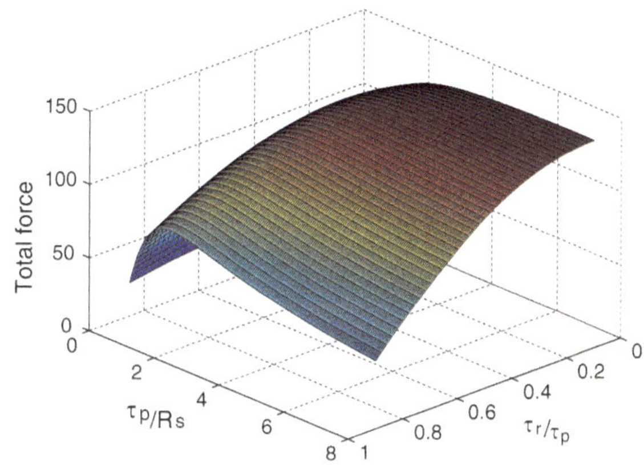

(a) Output force as function of τ_p/Rs and τ_r/τ_p

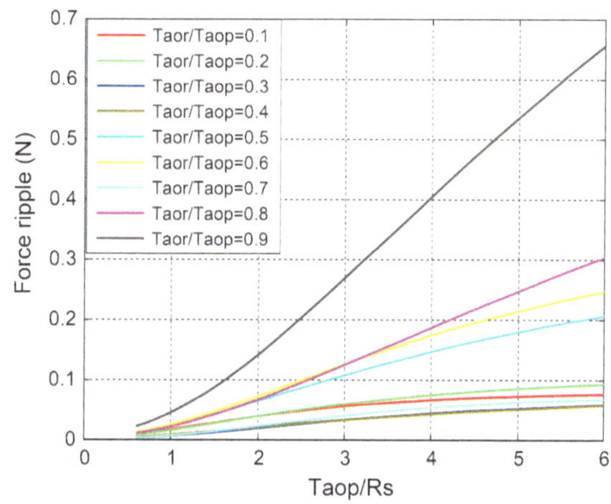

(b) Force ripple as function of τ_p/Rs and τ_r/τ_p

5.3.3 Determination of Independent Parameters

Among the five parameters, only k_{ra} is not closely related to the rest four parameters. The force output and force ripple with respect to k_{ra} are presented in Fig. 5.9. It is found that when k_{ra} arrives at the lowest value, the force output is maximized and the force ripple is minimized. A low ratio of k_{ra} represents a relative increase of internal PM length, and thus high force output. However, large volume of magnets may increase the system cost significantly. In this section, k_{ra} is chosen to be 0.35, and thus $R_r = 7$ mm for assembly convenience.

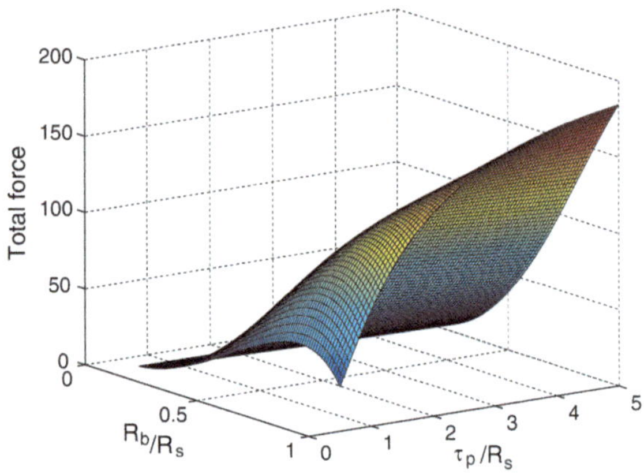

(a) Output force as function of τ_p/Rs and τ_r/τ_p

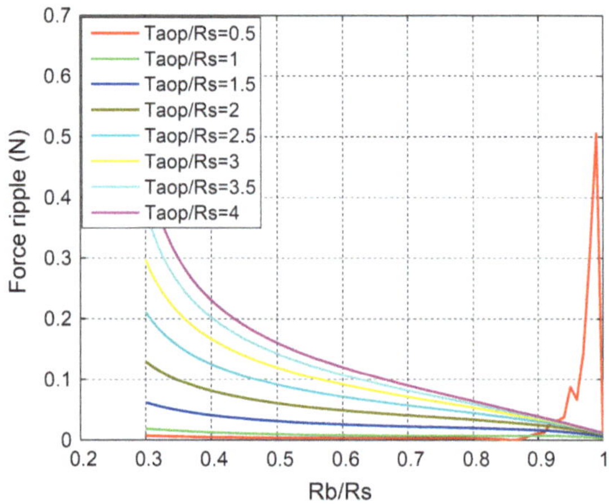

(b) Force ripple as function of τ_p/Rs and τ_r/τ_p

Fig. 5.11 Force output and force ripple versus k_{ps} and k_{bs}, at $k_{ab} = 0.6$, $k_{rp} = 0.6$, $k_{ra} = 0.35$

5.3.4 Determination of Dependent Parameters

It is known that k_{bs} is closely related to k_{ab} and k_{ps}, and k_{ps} is dependent on k_{rp}. The variation of force output and force ripple with respect to k_{rp} and k_{ps} are illustrated in Fig. 5.10. The force ripple grows rapidly when k_{ps} is larger than 3.0, but the force output only rises slightly. Therefore, the value of k_{ps} can be set to 3.0 or less. It shows that when $k_{ps} = 3.0$ and $k_{rp} = 0.3$, the force output approaches to the peak value and the force ripple is less than 5 %. Therefore, $k_{ps} = 3.0$ and $k_{rp} = 0.3$ are selected for the machine design. To analyze the interdependence of k_{ps} and k_{bs}, the variations of force output and force ripple with respect to k_{ps} and k_{bs} are plotted in Fig. 5.11. From this figure, it is found that when $k_{ps} = 3.0$, $k_{bs} = 0.85$, the force output is very close to the peak point and the force ripple is less than 5 %. Similarly, to investigate the

Fig. 5.12 Force output and force ripple versus k_{bs} and k_{ab}, at $k_{ra} = 0.35$, $k_{ps} = 3$, $k_{rp} = 0.3$

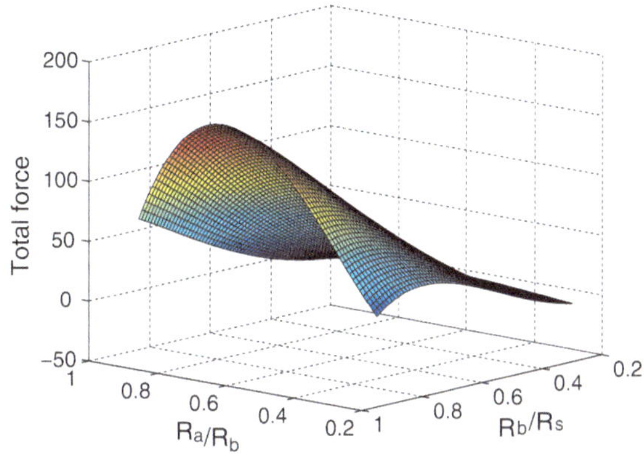

(a) Output force as function of τ_p/R_s and R_b/R_s

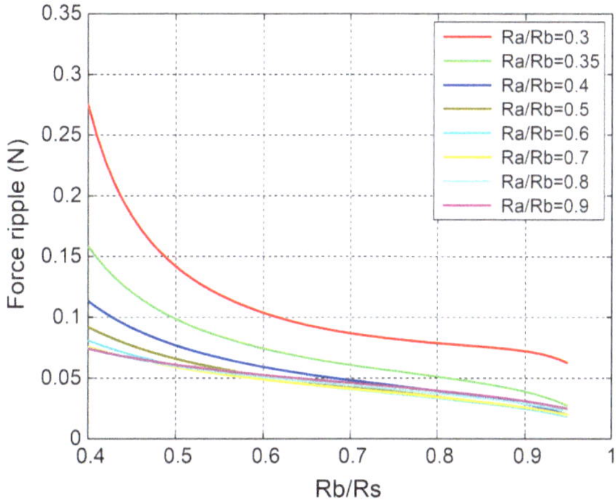

(b) Force ripple as function of τ_p/R_s and R_b/R_s

dependence of k_{bs} and k_{ab}, the variations of force output and force ripple with respect to k_{bs} and k_{ab} are plotted in Fig. 5.12. According to the figure, when $k_{bs} = 0.85$ and $k_{ab} = 0.6$, the force output gains the peak value, and the force ripple is less than 5 %. Therefore, the value of k_{ps} is set to 3.0, k_{rp} to 0.3, k_{bs} to 0.85, and k_{ab} to 0.6 for the design of the PM tubular linear machine.

The analytical models can provide a useful tool for assessing the influence of structure parameters of linear machines on their output performance. From the obtained dimensionless parameters and the maximum radius, R_s, in the last section, the final values for the structure parameters of the tubular linear machine can be obtained. The result is presented in Table 5.5.

Table 5.5 Final values of structure parameters

Outer radius of external Halbach array R_s	40 (mm)
Inner radius of external Halbach array R_b	34 (mm)
Outer radius of internal Halbach array R_a	20 (mm)
Inner radius of internal Halbach array R_r	7 (mm)
Magnetic pole pitch τ_p	120 (mm)
Axial width of radially magnetized PMs τ_r	36 (mm)

5.4 Design of Back Iron

Back irons are assumed to be infinitely permeable, and thus they can affect magnetic path greatly without magnetic saturation. The structures of tubular linear machines with different back iron patterns are illustrated in Fig. 5.13. The major difference among the four figures in Fig. 5.13 lies in the arrangement of back irons. Specifically, Fig. 5.13a shows back irons mounted on both inner and outer layers of PMs, Fig. 5.13b shows single layer of back iron inside the inner PMs, Fig. 5.13c single layer of back iron outside the outer PMs, and Fig. 5.13d without any back iron. In the following sections, we will study the effect of back irons on the magnetic field and force output of linear machines with dual Halbach arrays.

5.4.1 Magnetic Circuits of Four Patterns

Magnetic circuit presents the magnetic field distribution in electromagnetic machines intuitively, and is useful for the analysis of back iron effect on magnetic field. The magnetic circuits of four back iron patterns are shown in Fig. 5.14.

Figure 5.14a indicates that the flux lines in the radial magnets flows in the radial direction, whereas the flux lines in the axial magnets are compressed to the inner side of the machine. Therefore, flux lines in the radial magnets are straight, while flux lines in the axial magnets are curved due to the influence of back irons. It is noted that the axial flux component in the winding area is very small, and the radial one is enhanced greatly. This can help to increase the axial force output of the linear machines. Assume that back irons are infinitely permeable. The magnetic resistance is ignorable, which can improve the working efficiency of the machine significantly.

5.4.2 Magnetic Field Variation in 3D Space

In this section, we will study the back iron effect on the magnetic field distribution in 3D space of the linear machines with dual Halbach arrays. Magnetic field distributions of four back iron patterns are analyzed and compared to illustrate the

Fig. 5.13 Four types of back iron patterns for dual Halbach array

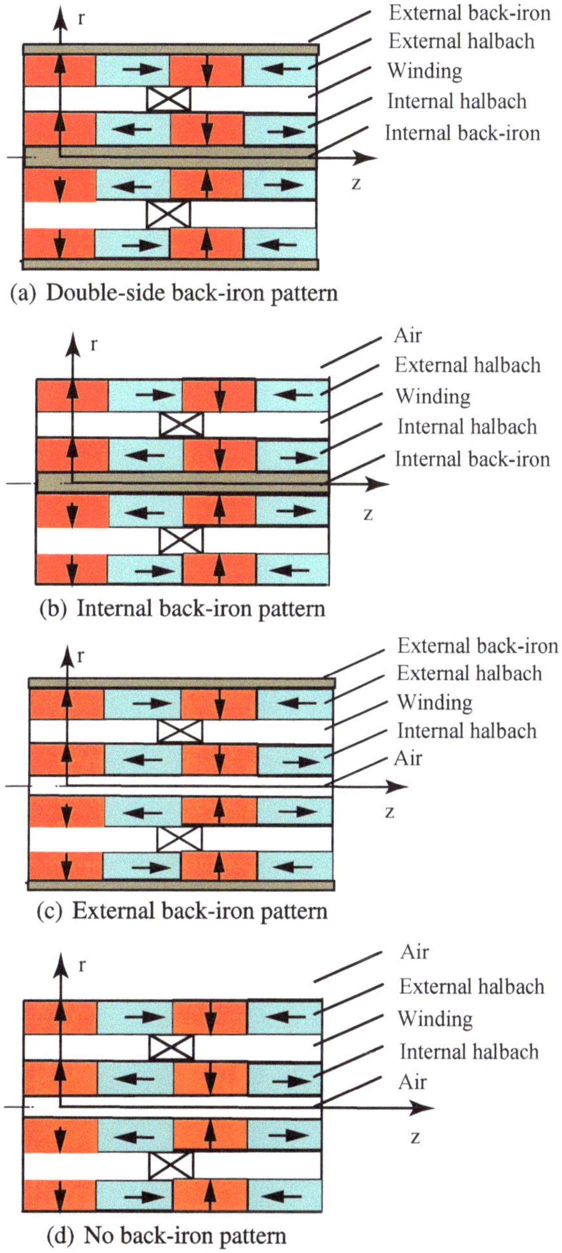

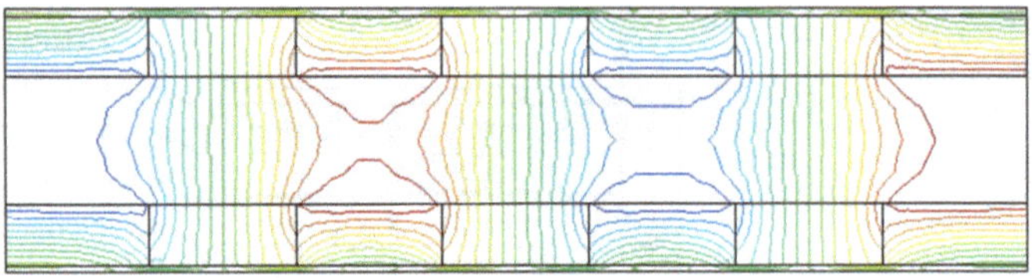

(a) Double-sided back-iron pattern

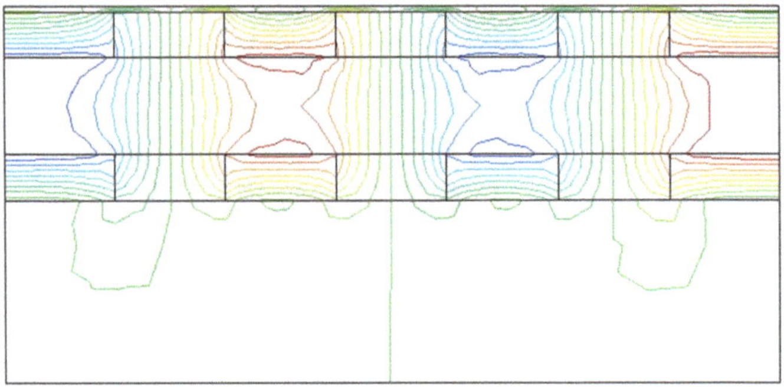

(b) External back-iron pattern

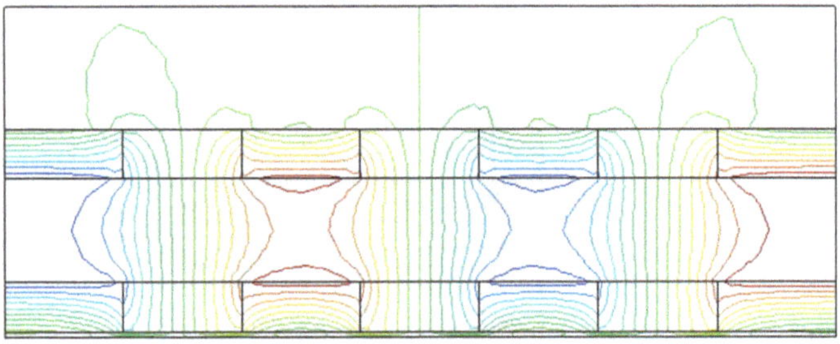

(c) Internal back-iron pattern

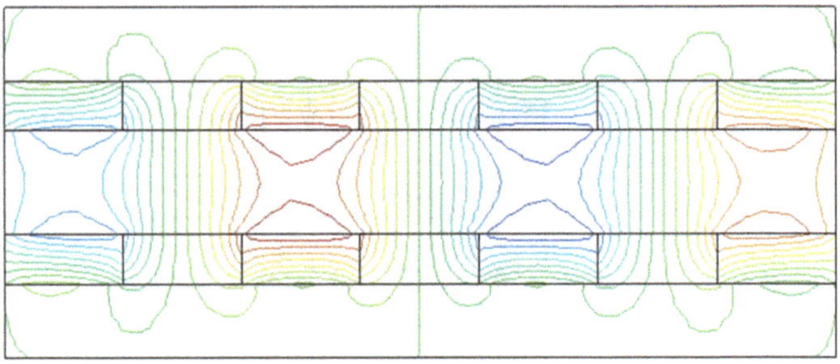

(d) Non back-iron pattern

Fig. 5.14 Flux paths for four types of back iron patterns

Table 5.6 Structure parameters of the tubular linear machine

Outer radius of external Halbach array R_s	40 (mm)
Inner radius of external Halbach array R_b	32 (mm)
Outer radius of internal Halbach array R_a	15 (mm)
Inner radius of internal Halbach array R_r	7 (mm)
Pole pitch τ_p	40 (mm)
Axial width of radially magnetized PMs τ_r	20 (mm)

influence of back irons. The structure parameters of the linear machine are shown in Table 5.6.

5.4.3 *Overall Field Distribution of Four Patterns*

Because permanent magnets in the linear machine are arranged periodically, magnetic field always varies in a period of double pole pitch if ignoring longitudinal end effect. Overall field distributions of four back iron patterns in one period are simulated with numerical method and presented in 3D coordinate system. As the magnetic field is axially symmetric, only r and z are concerned. Field distribution for double-sided back iron pattern, external back iron pattern, internal back iron pattern, and no back iron pattern is shown in Figs. 5.15, 5.16, 5.17, and 5.18 respectively. In each of these figures, the variation of both flux components in radial direction B_r and in axial direction B_z is presented with respect to r and z. The four figures show that magnetic field distributions of the four patterns share a similar variation tendency. The variation of B_r and B_z in the magnetic areas is consistent with the polarization pattern of magnets. The variation of B_r is continuous in radial direction, while the variation of B_z is not smooth at the boundary of different materials. It is because the normal component of flux density is continuous at the boundary of two neighboring media with different permeabilities. It is also found that magnetic field distribution is distinctive near internal back iron among the four back iron patterns. Internal back irons in Figs. 5.15 and 5.17 improve the radial flux fields significantly. The peak values exceed 1.0T. The maximum radial flux density in motor designs without internal back iron is less than 1.0T.

5.4.4 *Comparative Study of Field Distributions of Four Patterns*

5.4.4.1 Field Variation Versus Axial Distance *z*

Field variation of four back iron patterns versus axial distance z at the center of winding region, i.e., $r = (R_a + R_b)/2$, is shown in Fig. 5.19. It is found that both B_r and B_z vary in sinusoidal form, and B_r is even-symmetric about $z = 0\,\text{mm}$, B_z is

Fig. 5.15 Flux field for double-sided back iron pattern with dual Halbach array

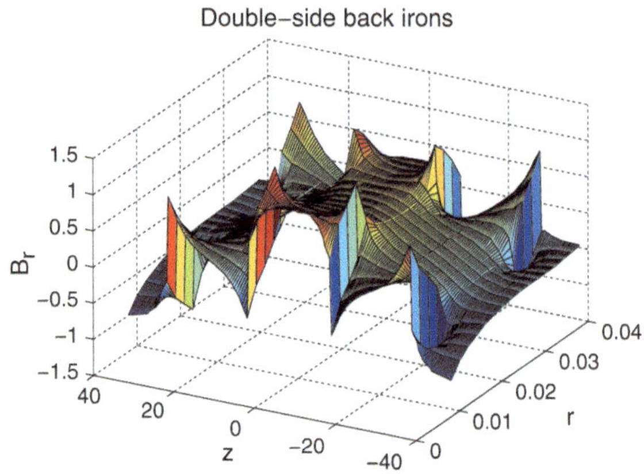

(a) Variation of radial flux component B_r

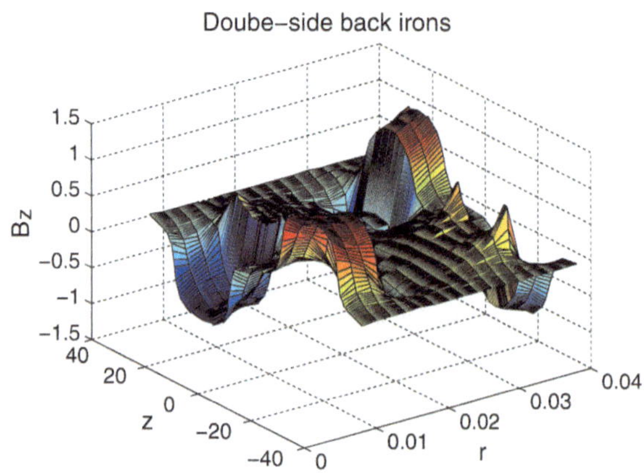

(b) Variation of axial flux component B_z

odd-symmetric. Figure 5.19a shows that double-sided back iron pattern and internal back iron pattern can generate higher radial magnetic flux field than the other two patterns, which means that the internal back iron effect is dominant. Figure 5.19b shows that the employment of back irons, especially internal back iron, does decrease axial field B_z. Therefore, we may conclude that the employment of back irons, especially the internal back iron, can improve the radial field B_r and weaken the axial field B_z in the winding area. This helps to increase the axial force output of tubular linear motors with Halbach array.

Magnetic field variation for four back iron patterns versus axial distance z at the center of internal magnet area, i.e., $r = (R_r + R_a)/2$, is shown in Fig. 5.20. It is found that the variation of B_r is not smooth in the z direction, while the variation of B_z is continuous. Peak points of B_r always appear at the axial boundary of magnets. It

Fig. 5.16 Magnetic field for linear motors with external back iron pattern

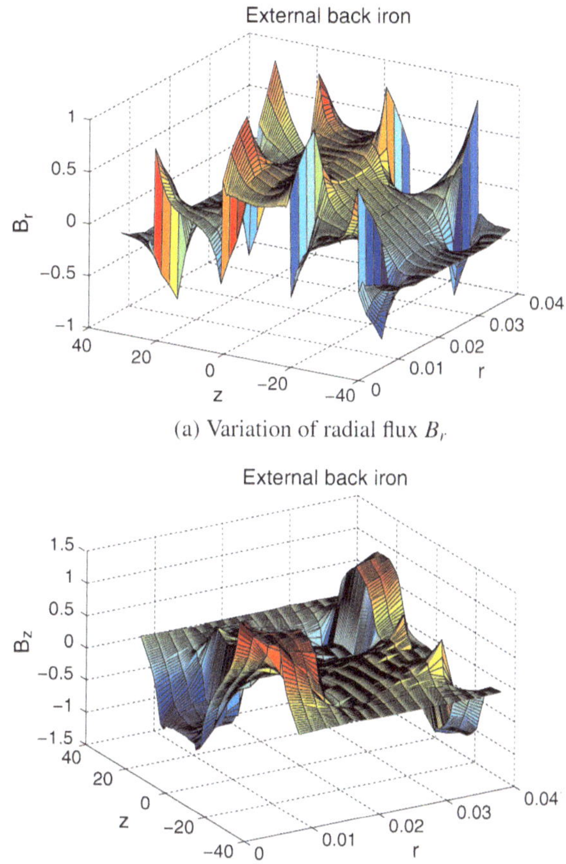

(a) Variation of radial flux B_r

(b) Variation of axial flux B_z

is due to the edge effect of radially magnetized PMs. In the internal magnet area, internal back iron is dominantly effective. It is found that the effect of internal back iron is to increase radial field component B_r at the expense of a slight drop of axial field component B_z. Therefore, the internal back iron can induce magnetic path in the radial direction. Magnetic field variation for four back iron patterns versus axial distance z at the center of external magnet area, i.e., $r = (R_b + R_s)/2$, is shown in Fig. 5.21. In this area, the influence of external back iron on the magnetic field is more effective. It is found that the external back iron can increase both B_r and B_z. It is because the external back iron can reduce magnetic energy loss and increase magnetic flux.

Fig. 5.17 Magnetic field for linear motors with internal back iron pattern

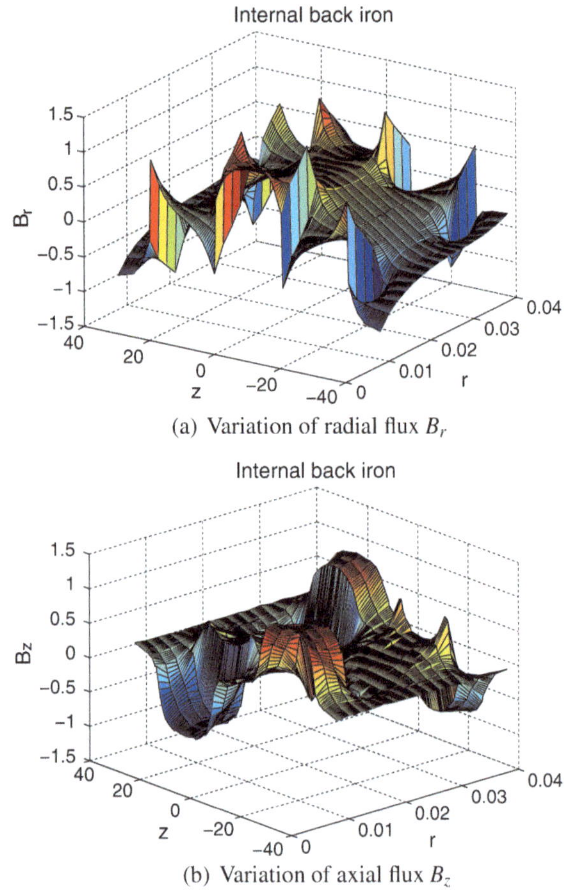

5.4.4.2 Field Variation Versus Radial Distance *r*

Magnetic field variation versus radial distance r at the center of a radial magnet, $z = 0\,\text{mm}$, is illustrated in Fig. 5.22. Since axially magnetized PMs in four back iron patterns are all oddly symmetric about $z = 0\,\text{mm}$, B_z at this position is nearly equal to zero as indicated in Fig. 5.22b. Figure 5.22a shows that back irons affect the nearby field greatly. It is found that the internal back iron is much more effective than external back iron in improving B_r. As shown in Fig. 5.22a, the internal back iron increases nearby field density form 0.225T to 1.2T, while the external back iron only increases nearby field density from 0.16 T to 0.4T. There are two reasons: 1. The internal back iron can save more magnetic energy; 2. The internal area crossed by magnetic flux is much smaller. Magnetic energy saved by back irons is wasted in

Fig. 5.18 Magnetic field for linear motors with no back iron pattern

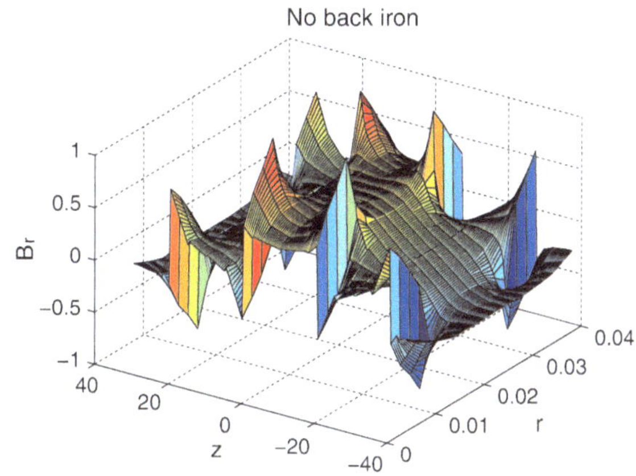

(a) Variation of radial flux B_r

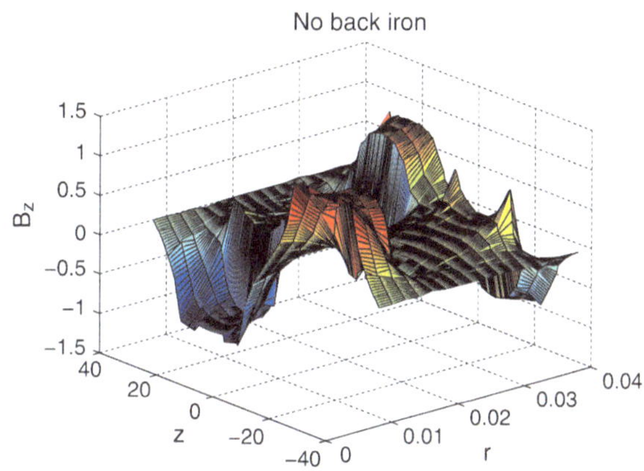

(b) Variation of axial flux B_z

the air area on the outside of two layers of PMs as shown in Fig. 5.14. Because of the linear magnetic material property of air, the relationship between magnetic energy and flux density can be defined as follows.

$$W = \frac{1}{2}\int_V B \bullet H dv = \frac{1}{2}\int_V \frac{B^2}{\mu_0} dv. \tag{5.5}$$

From Eq. (5.5), it is known that the magnetic energy is directly proportional to B^2. Flux densities versus radial distance, r, at internal and external air space is shown in Fig. 5.23. It is found that the flux density in the internal air space is greater than that of external air area. Therefore, the internal back iron can reduce more magnetic

Fig. 5.19 Field variation at the center of winding region

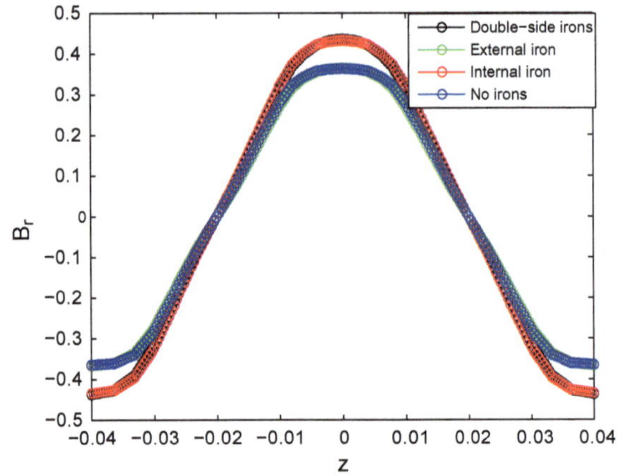

(a) Variation of B_r for four back-iron patterns

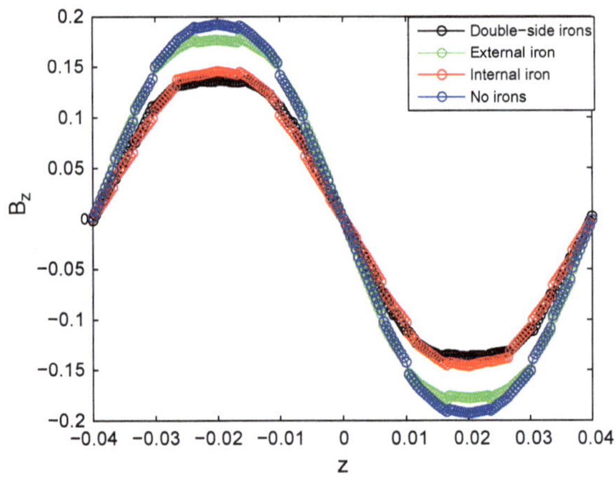

(b) Variation of B_z for four back-iron patterns

energy loss. The saved magnetic energy can generate extra magnetic flux. The area crossed by magnetic flux increases versus radial distance, and thus the back iron effect decreases accordingly.

5.4.5 Force Variation

Force outputs with an input current density 2.0×10^6 A/m^2 of four back iron patterns are shown in Fig. 5.24. It is found the internal back iron is more effective and the external back iron has less effect on the force output. This result is consistent with the magnetic field analysis. Force output generated by dual Halbach array varies sinusoidally in axial direction, which facilitates the control implementation.

Fig. 5.20 Field variation at the center of internal magnet area

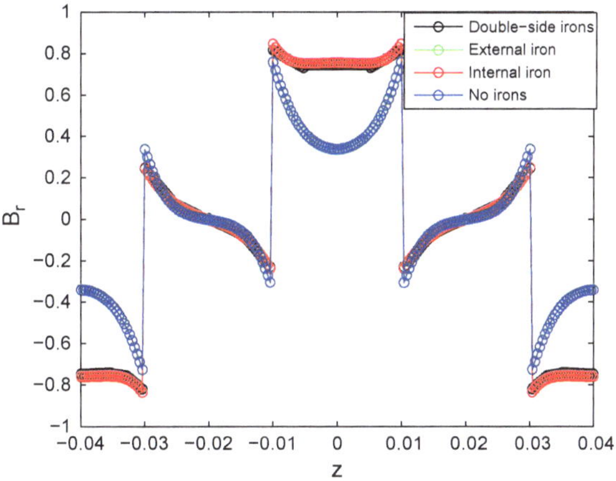

(a) Variation of B_r for four back-iron patterns

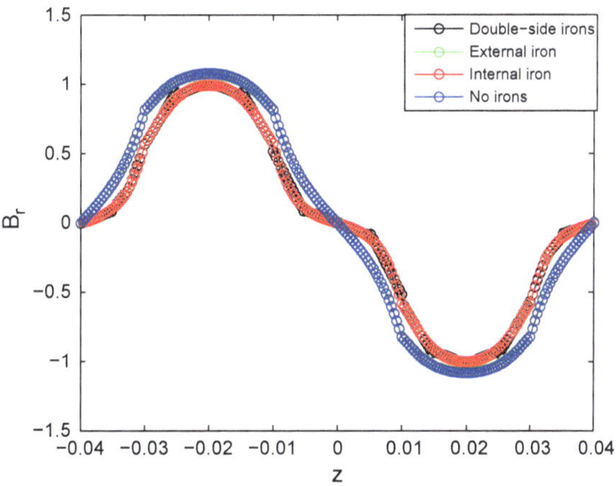

(b) Variation of B_z for four back-iron patterns

5.5 Conclusion

Based on the established analytical model, the design optimization with penalty method and scanning method is conducted to maximize the force output. The analytical models provide a useful tool for assessing the influence of structure parameters of linear machines on their output performance. Based on the parameter analysis, one set of values are selected for the structure design of the tubular linear machine with dual Halbach array. Furthermore, this chapter studies back iron effect on both magnetic field distribution and force output of a tubular linear motor with dual Halbach array in Ansoft environment. Numerical method is employed to simulate the magnetic field and force output of tubular linear machines with various type of back irons. Simulation result shows that internal back iron does help to enhance the

Fig. 5.21 Field variation at the center of external magnet area

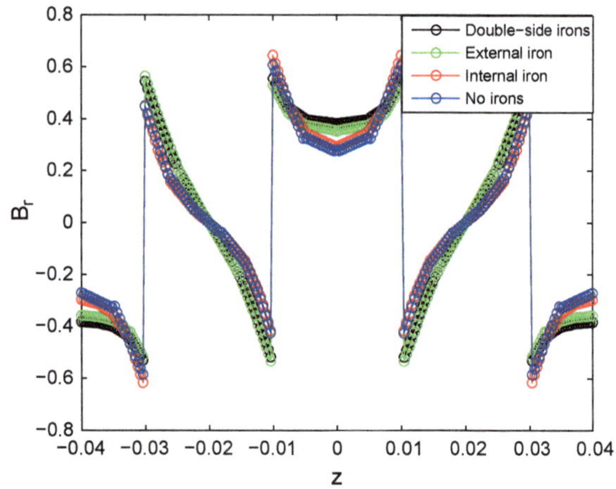

(a) B_r variation of four patterns

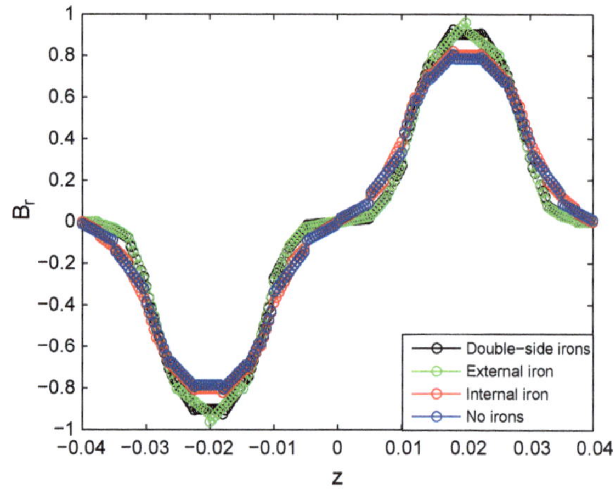

(b) B_z variation of four patterns

magnetic field flux density and improve the force output, while external back iron has little effect on them. The back irons have two important functions. One is to reduce the magnetic energy loss and the other is to enhance the magnetic flux density in the radial direction. In addition, the effectiveness of the back iron is related to the magnetic field on one side of back irons since the improvement of magnetic energy efficiency caused by the back irons depends on magnetic field. The study results can be employed in the magnetic circuit optimization, and design and analysis of electromagnetic machines.

Fig. 5.22 Magnetic field variation at $z = 0\,\text{mm}$

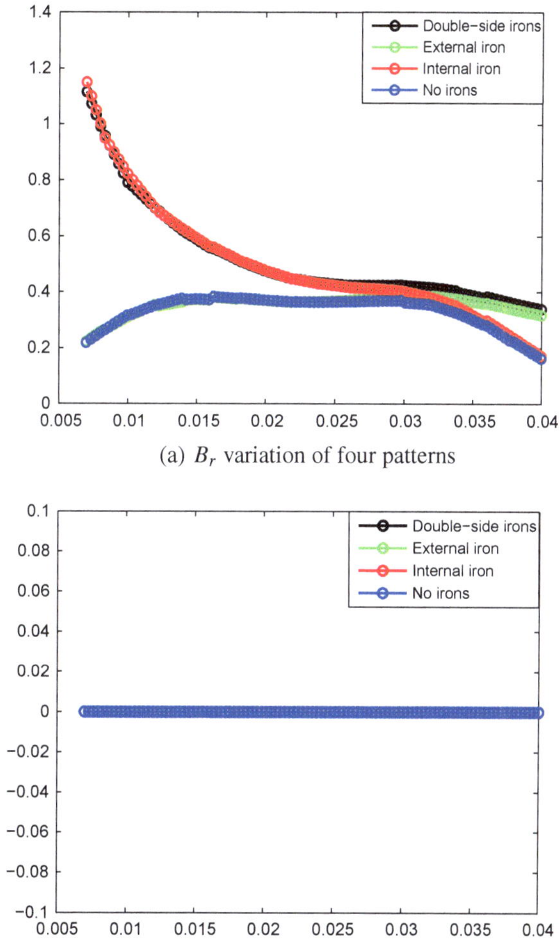

(a) B_r variation of four patterns

(b) B_z variation of four patterns

Fig. 5.23 Flux density of air area at internal and external air space

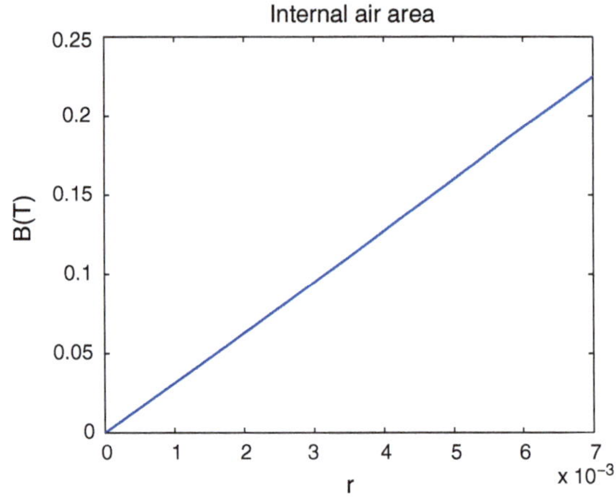

(a) Flux density of internal air area

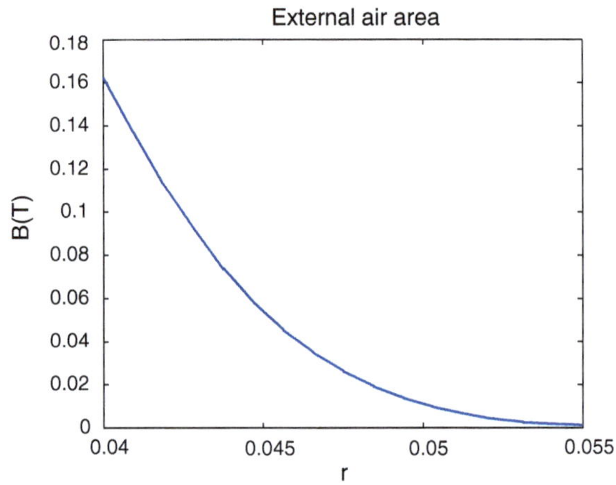

(b) Flux density of external air area

Fig. 5.24 Force variation of four back iron patterns

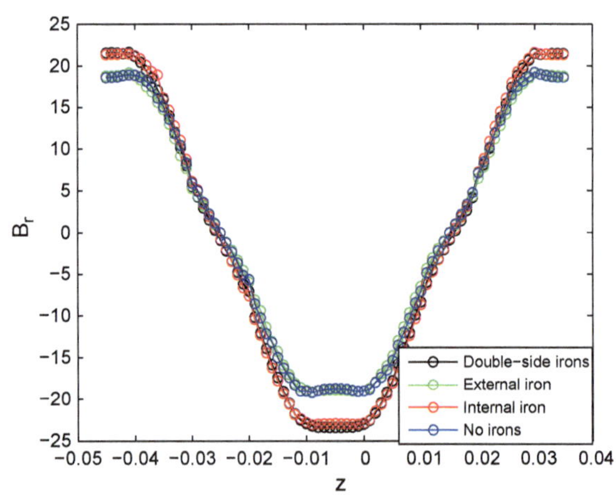

References

1. Jie W, Suqin W (1999) Research in overall optimization method and application of motor optimization design. Prog Electromagn Res 36(6):15–17
2. Widdowson GP, Howe D, Evison PR (1991) Computer-aided optimization of rare-earth permanent magnet actuators. In: International conference on computation in electromagnetics, 1991. IET, London, UK, pp 93–96, 25–27 November 1991
3. Jang SM, Lee SH, Cho HW et al (2003) Design and analysis of helical motion permanent magnet motor with cylindrical Halbach array. IEEE Trans Magn 39(5):3007–3009
4. Wang J, Howe D (2005) Tubular modular permanent-magnet machines equipped with quasi-Halbach magnetized magnets-part I: magnetic field distribution, EMF, and thrust force. IEEE Trans Magn 41(9):2470–2478
5. Zhao L, Peng Y, Sha C et al (2008) Cylindrical linear halbach arrays for permanent magnet tubular linear machines. In: International conference on electrical machines and systems, 2008. ICEMS 2008. IEEE, Wuhan, China, pp 175–178, 17–20 October 2008
6. Li L, Xuzhen H, Donghua P et al (2011) Magnetic field of a tubular linear motor with special permanent magnet. IEEE Trans Plasma Sci 39(1):83–86
7. Zhu Y, Lee S, Cho Y (2010) Optimal design of slotted iron core type permanent magnet linear synchronous motor for ropeless elevator system. In: 2010 IEEE international symposium on industrial electronics, Bari, Italy, pp 1402–1407, 4–7 July 2010
8. Atallah K, Howe D, Mellor PH (1997) Design and analysis of multi-pole Halbach (self-shielding) cylinder brushless permanent magnet machines. In: 1997 Eighth international conference on electrical machines and drives (Conf. Publ. No. 444). IET, Cambridge, UK, pp 376–380, 1–3 September 1997
9. Chung MJ, Lee MG, Lee SQ et al (2001) Optimal design and development of linear brushless permanent magnet motor. In: IEEE international: electric machines and drives conference, 2001. IEMDC 2001. Cambridge, UK, pp 436–441, 17–20 June 2001
10. Yan L, Hu J, Jiao Z et al (2011) Flux distribution and back-iron influence of tubular linear motors. In: 2011 International conference on fluid power and mechatronics (FPM). IEEE, Beijing, China, pp 1–6, 17–20 August 2011
11. Yan L, Peng J, Jiao Z et al (2015) Flux field and thrust analysis of permanent-magnet linear machines with isolated movers. IEEE Trans Magn 51(8):1–8
12. Yan L, Hu J, Yao N et al (2012) Flux field formulation and back-iron analysis of tubular linear machines. IEEE Trans Magn 48(10):2617–2626
13. Wang J, Jewell GW, Howe D (2001) Design optimisation and comparison of tubular permanent magnet machine topologies. IEE Proc Electr Power Appl 185(5):456–464
14. Yan L, Li W, Jiao Z et al (2015) Novel tubular switched reluctance motor with double excitation windings: design, modeling, and experiments. Rev Sci Instrum 86(12):125004
15. Wang T, Yan L, Jiao Z et al (2015) Analytical modeling of linear oscillating motor with a mixed method considering saturation effect. Sens Actuators A: Phys 234:375–383

Chapter 6
Research Prototype and Experiments

Abstract In this chapter, the experimental investigation is conducted on the research prototype of the tubular linear machine with dual Halbach array to verify the analytical magnetic field model, force output model, and armature reaction field model proposed in Chaps. 2, 3, and 4, respectively. The experimental results also facilitate the visualization of the magnetic field distribution, the force output variation, and armature reaction field analysis of the linear machine. As shown in Fig. 6.1, the experimental investigation of the tubular linear machine consists of three major tasks: measurement of the magnetic field distribution, the force output variation, and the armature reaction change. This chapter begins with description of research prototype, magnetic field measurement procedure of the tubular linear machine, and the corresponding data processing and analysis. The force and the armature reaction measurements are carried out on the machine and compared with the theoretical models. The experimental results confirm the proposed analytical model for further control study of the linear machine.

6.1 Research Prototype

A tubular linear machine with dual Halbach array is developed for experimental investigation as shown in Fig. 6.2. The windings are mounted on the mover that in turn is fixed on the two guiders. The guiders can slide back and forth on the linear bearing. The bearings are installed on the covers of the stator on which the PM array is mounted. The structure of the research prototype is shown in Fig. 6.3 and design parameters are given in Table 6.1. The maximum linear stroke is 36 mm. PMs in the machine are sintered NdFeB35 with $B_{rem} = 1.2$T and $\mu_r = 1.0997$. The radially magnetized PMs are replaced with segments of diametrically magnetized sector PMs for the convenience of manufacturing and cost reduction.

L. Yan et al., *Electromagnetic Linear Machines with Dual Halbach Array*, DOI 10.1007/978-981-10-2309-5_6

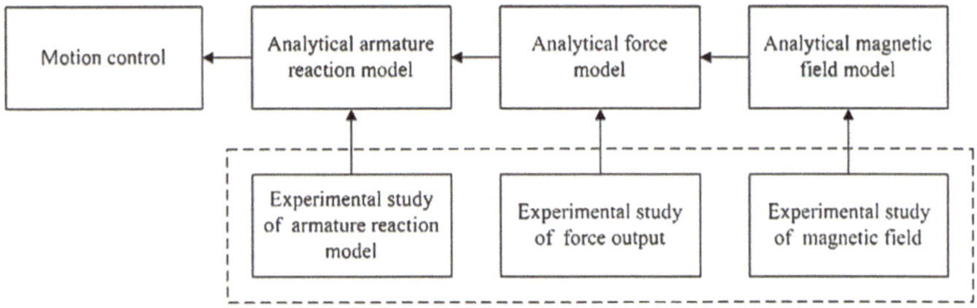

Fig. 6.1 Experimental investigation on tubular linear machine

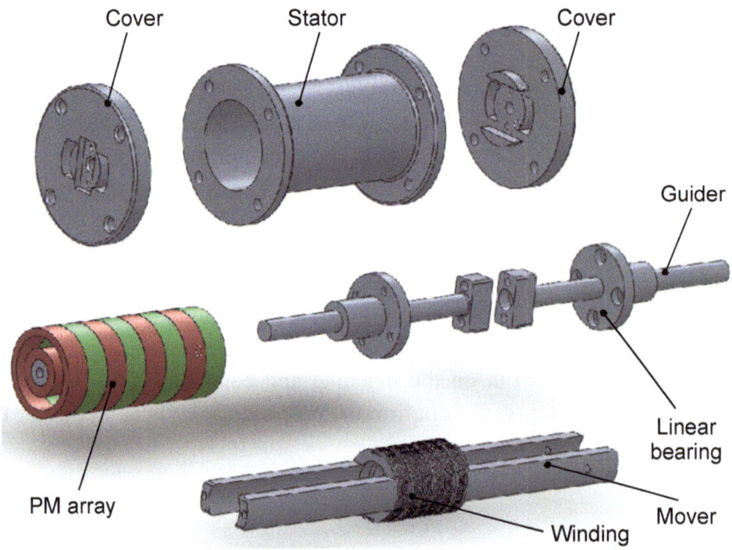

(a) 3D exploded view

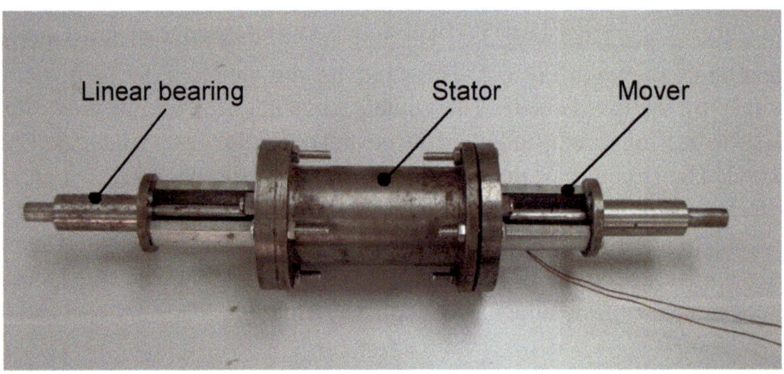

(b) Research prototype

Fig. 6.2 Linear machines with dual Halbach array

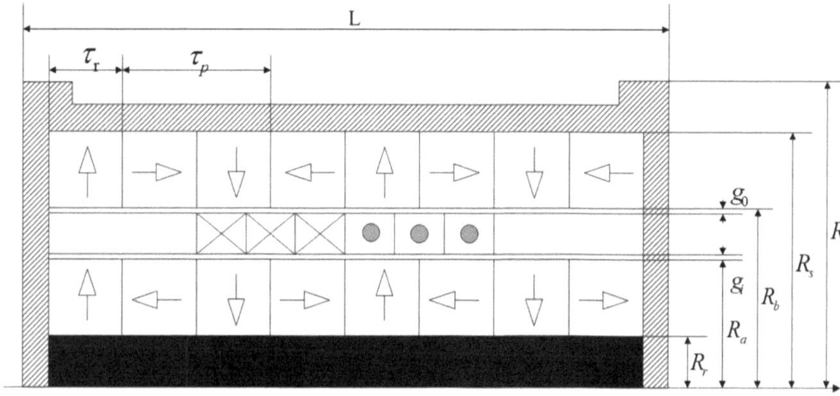

Fig. 6.3 Structure of the research prototype

Table 6.1 Design parameters of research prototype

Maximum radius R_o	30 mm	Outer rad of ext PM R_s	17 mm
Machine length L	89 mm	Inner rad of ext PM R_b	14 mm
Width of radial PM τ_r	9 mm	Outer rad of int PM R_a	9 mm
Pole pitch τ_p	18 mm	Inner rad of int PM R_r	5 mm
Number of poles n	4	Number of winding turns	100
Outer air gap g_o	0.2 mm	Inner air gap g_i	0.2 mm

6.2 Experimental Investigation on Magnetic Field

Experimental measurement on the magnetic flux density can be done readily in some special cases [1–4]. In recent years, linear motors have been developed by researchers to overcome the drawbacks of conventional linear mechanisms realized by several ball screws [5–8]. In this section, an experimental apparatus has been developed. The experimental works are conducted on the flux field distribution to validate the derived analytical models of magnetic field in previous chapter. The design parameters and material characteristics used in the numerical computation are consistent with those in experiments.

6.2.1 Experimental Apparatus of Magnetic Field Measurement

An experimental apparatus is developed for magnetic field measurement in 3D space as illustrated in Fig. 6.4. The research prototype is mounted on a platform of the

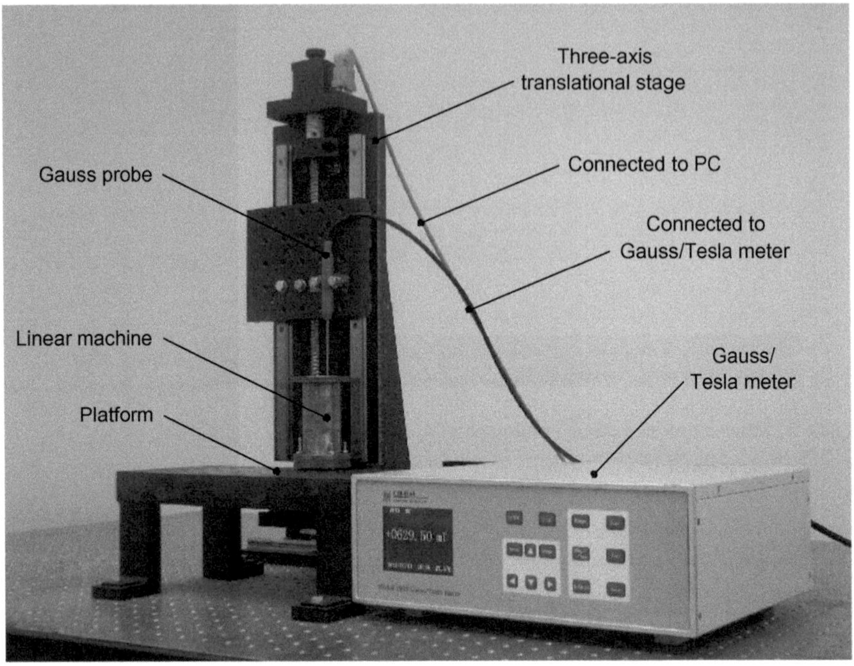

Fig. 6.4 Experimental testbed for magnetic field measurement

apparatus. A Gauss probe is installed on the end effect of a three-axis translational stage. Under the PC controller, the probe can pinpoint to any position inside the linear machine and measure the flux density. The measured data can be either displayed on the Gauss meter or transferred to PC [9–12].

6.2.2 *Magnetic Field Variation in the Radial Direction*

Figure 6.5 gives radial variation of the radial and axial flux density at constant axial distance $z = 0$ mm and rotation angle $\theta = 0$. Difference between analytical solutions and experimental results is less than 4 mT. Since the linear motor is symmetrically constructed and axial magnets are symmetrically arranged, axial field B_z at $z = 0$ mm is expected to zero. Figure 6.5b shows a tolerant deviation less than 3 mT between analytical solutions and experimental results which is largely due to manufacturing error or unbalanced magnetization. Figure 6.6 gives radial variation of the radial

Fig. 6.5 Magnetic field variation versus r at $z = 0\,\text{mm}$

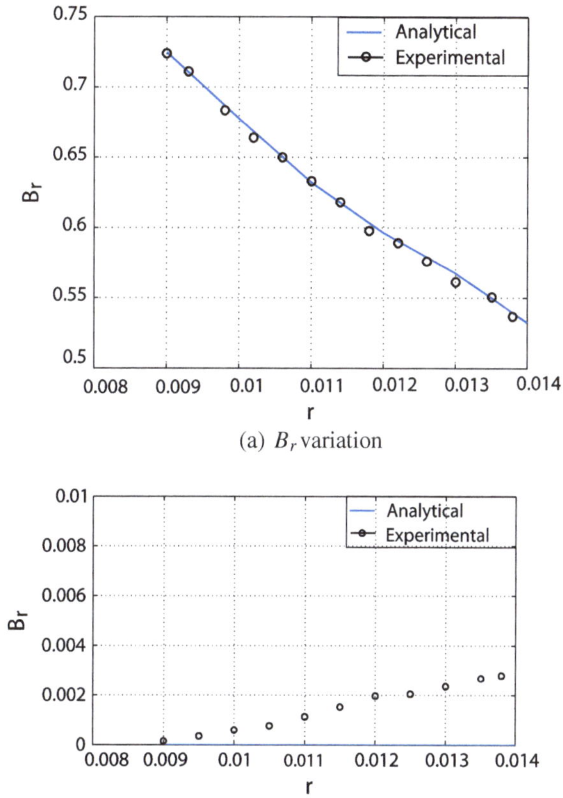

(a) B_r variation

(b) B_z variation

and axial flux density at constant axial distance $z = 9\,\text{mm}$ and rotation angle $\theta = 0$. Difference between analytical solutions and experimental results is less than 3 mT. B_r at $z = 9\,\text{mm}$, i.e. axial center of the radial magnet, is zero computed from analytical model of the infinite length motor. However, experiment results from the prototype deflect the analytical solutions mainly because radial magnets are not averagely placed on the sides.

6.2.3 Magnetic Field Variation in the Axial Direction

Magnetic field variations versus z at constant radius $r = 10.5, 11.5, 12, 12.5\,\text{mm}$ and $\theta = 0$ are given in Figs. 6.7, 6.8, 6.9 and 6.10, respectively. In this section, experimental results are measured near $z = 0\,\text{mm}$ to ignore longitude fringe effect.

Fig. 6.6 Magnetic field variation versus r at $z = 9\,\mathrm{mm}$

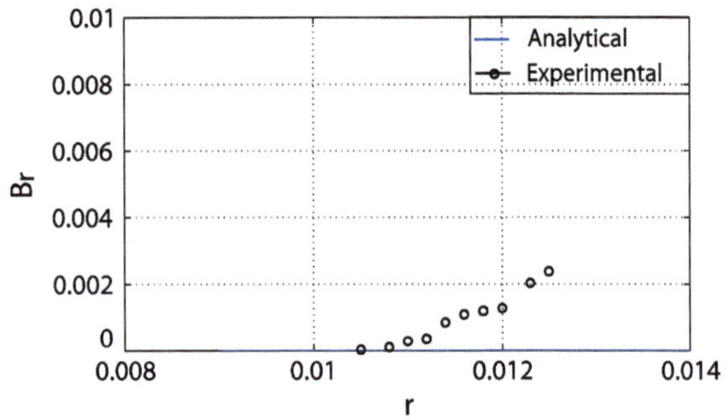

(a) B_r variation

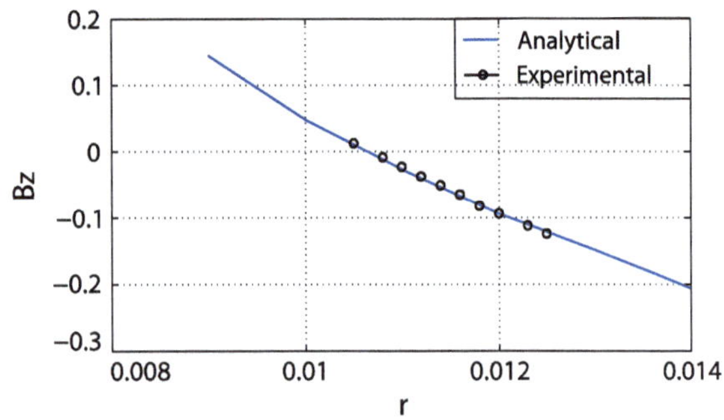

(b) B_z variation

It is found that the analytical solutions fit with the experimental results closely, which means the analytical model is accurate for magnetic field computation far from the machine end. Therefore, the assumption that the machine length is infinite utilized in establishing the analytical model is proved to be sound.

6.2.4 Magnetic Field Variation in the Circumferential Direction

Magnetic field variations in circles with radius $r = 12.5\,\mathrm{mm}$ at axial distances $z = 0\,\mathrm{mm}$ and $z = 9\,\mathrm{mm}$ are illustrated in Figs. 6.11 and 6.12, respectively. It is assumed that dual Halbach array is composed of ring magnets that are either magnetized in the radial direction or axial direction. Therefore, there is no circumferential component of magnetic field. Both figures show that the magnetic field fluctuation in a circle is less than 5 mT, which is mainly owe to fabrication tolerance or magnetization error of ring magnets.

Fig. 6.7 Magnetic field variation versus z at $r = 10.5\,\text{mm}$

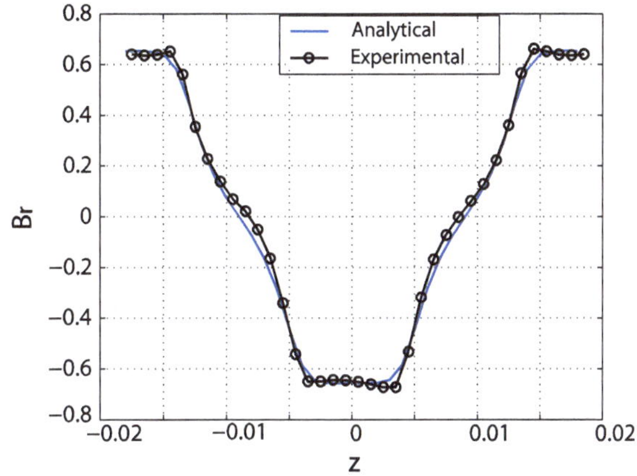

(a) B_r variation

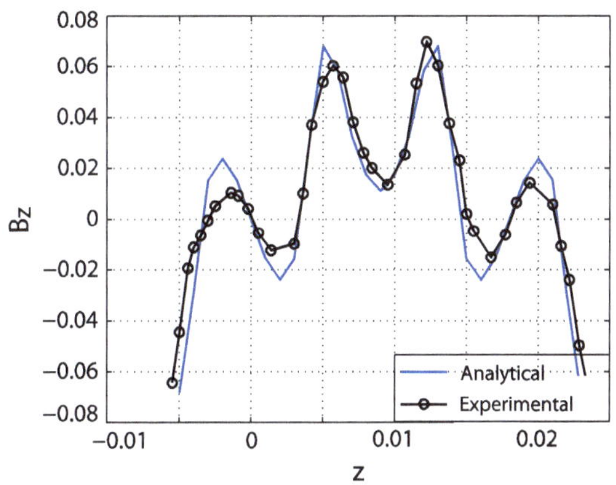

(b) B_z variation

6.2.5 Prediction of Machine Performance

Dual Halbach array consists of additional Halbach array with different magnetization pattern inside or outside conventional Halbach array, and magnetic field in the winding area is the superposition of two Halbach arrays [13–15]. In this section, magnetic field comparison among three magnet arrays is carried out, i.e., dual Halbach array, exterior Halbach array, and interior Halbach array. Magnetic field data

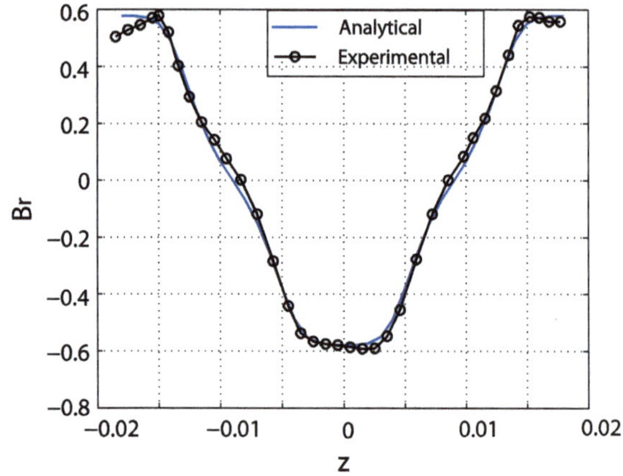

(a) B_r variation

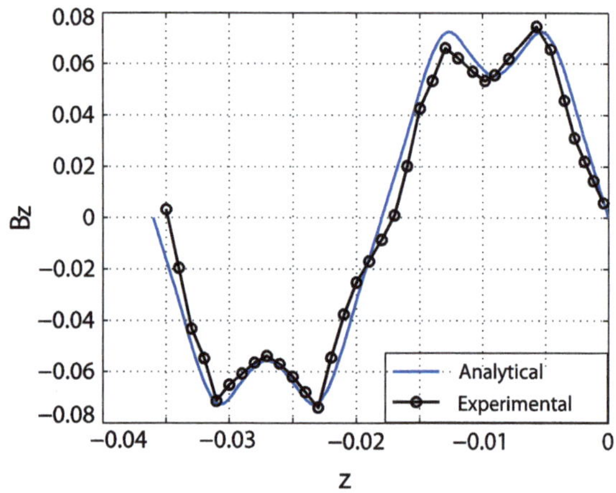

(b) B_z variation

Fig. 6.8 Magnetic field variation versus z at $r = 11.5\,\text{mm}$

in this section is obtained from numerical calculation in Ansoft environment, and the three machines with different magnet arrays have the same volume. Machine structures of interior and exterior Halbach arrays are shown in Fig. 6.13.

Figure 6.14 shows magnetic field distribution versus radial distance at $z = 0\,\text{mm}$. It is found that the radial flux field generated by dual Halbach array is much greater compared with that of other two topologies, and axial field is near to zero due to the approximately symmetrical structure.

Fig. 6.9 Magnetic field variation versus z at $r = 12\,\mathrm{mm}$

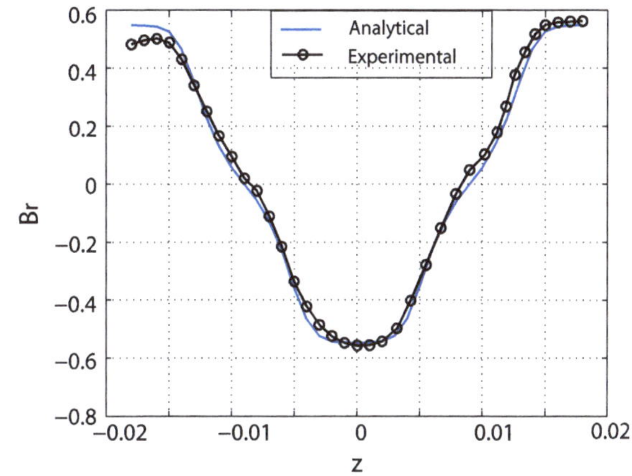

(a) B_r variation

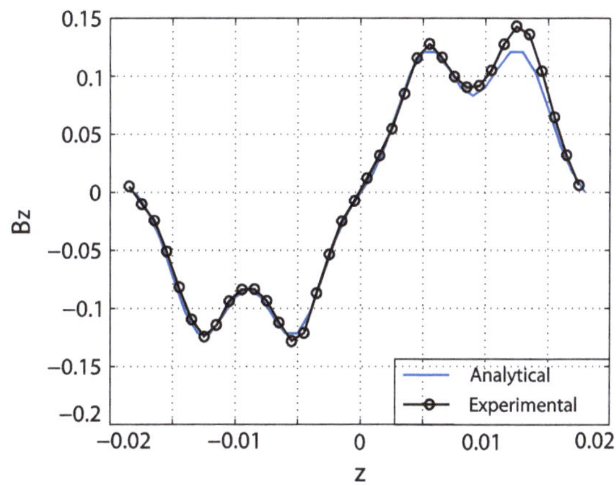

(b) B_z variation

Figure 6.15 shows magnetic field distributions versus axial distance z at the center of the winding area, i.e., $r = (R_a + R_b)/2$. It is found that the radial flux field in this area varies sinusoidally. The radial flux field of the dual Halbach array that contributes to force generation is superior to that of other two topologies. The axial flux field component of the dual Halbach array generated by two layers of PM poles is approximately equal to algebraic sum of that of other two topologies. Since axial fields created by interior and exterior Halbach array are in the opposite direction, the dual Halbach array can cancel the axial field, nearly equal to zero if the structure parameters are designed properly.

Fig. 6.10 Magnetic field variation versus z at $r = 12.5\,\text{mm}$

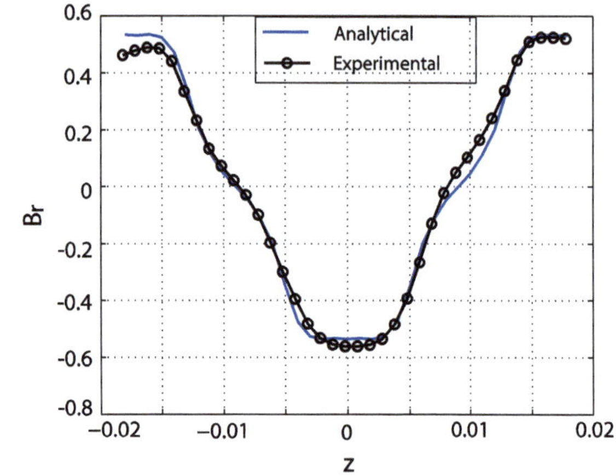

(a) B_r variation

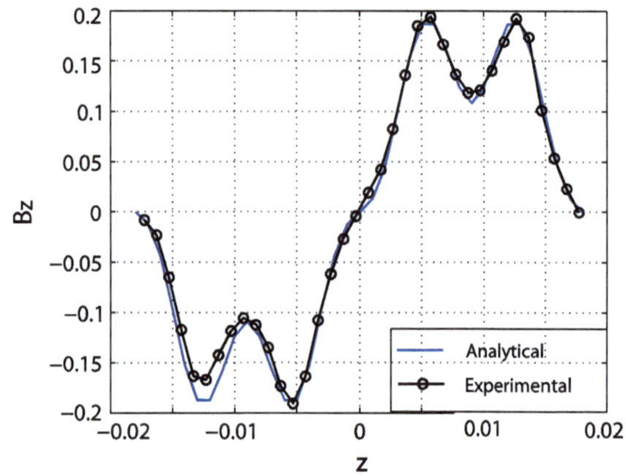

(b) B_z variation

Figure 6.16 describes force variation of three magnet arrays as a function z. It is found that force output of the linear machine with dual Halbach array approaches to twice of that of interior Halbach array and more than 30 % higher than that of exterior Halbach array. Both magnetic field and force comparisons show that linear machines with dual Halbach array have a better performance. Therefore, the dual Halbach array is promising to be applied in high power density systems like direct drive valve.

Fig. 6.11 Magnetic field variation versus θ at $r = 12.5\,\text{mm}$, $z = 0\,\text{mm}$

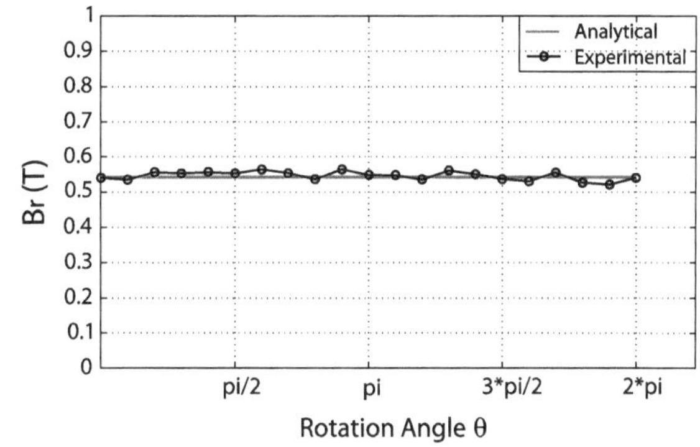

(a) B_r variation

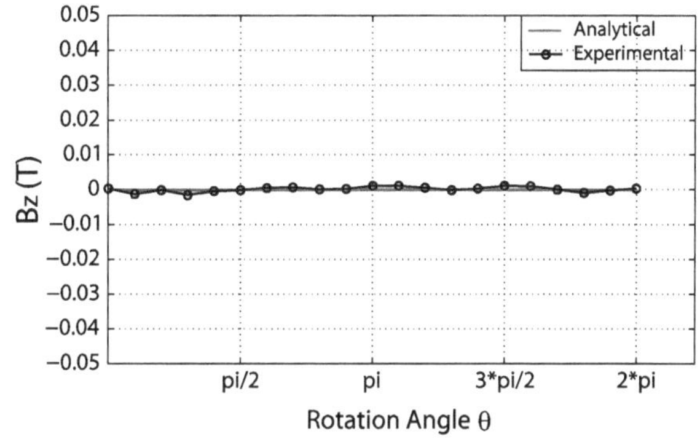

(b) B_z variation

6.3 Experiments on Force Output

6.3.1 Measuring Platform of Force Output

The measurement of force output is an essential part to analyze system performance [16–18]. In order to measure the force output, one measurement platform is developed as shown in Fig. 6.17. The strain type pressure sensor is mounted on output shaft of the mover. It is fixed on a high-precision translational motion stage. The mover can thus be adjusted to measure the force output for different mover positions.

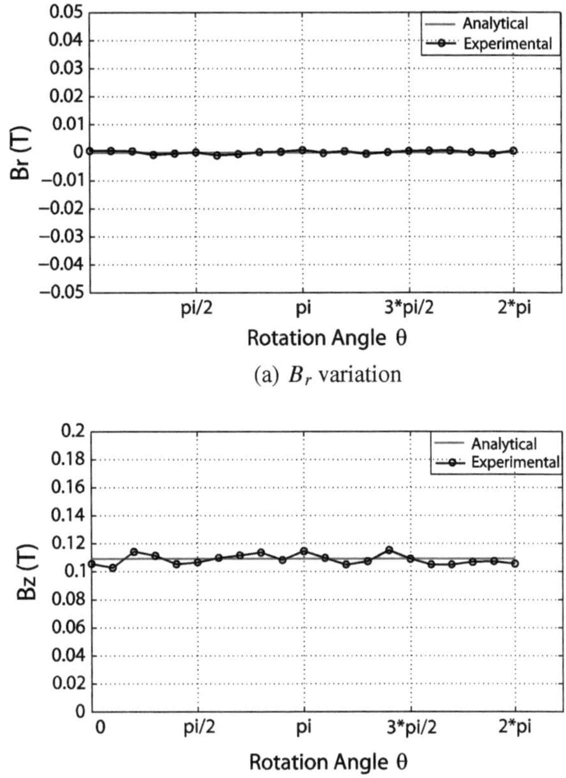

(a) B_r variation

(b) B_z variation

Fig. 6.12 Magnetic field variation versus θ at $r = 12.5\,\text{mm}$, $z = 20\,\text{mm}$

6.3.2 *Experiments on Force Output and Comparison*

The force output is measured within stroke and compared with the theoretical torque model with step size of 1 mm. The measurement result is shown in Fig. 6.18. It is found that the experimental result fits with the analytical model well, which indicates that the precision of the analytical force model is acceptable. The deviation between the experimental result and analytical model is caused by manufacturing error and end effect.

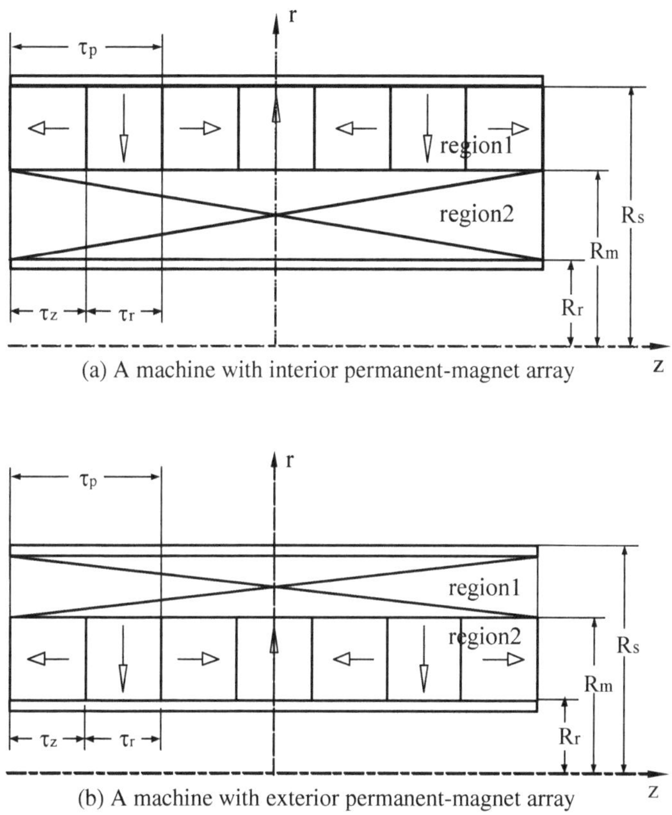

Fig. 6.13 Tubular linear machines with Halbach array

6.4 Experimental Investigation on Armature Reaction Field

The challenge for experimental study of the armature reaction field of the dual Halbach linear machine lies in the complexity of measurement of the armature field distribution [19–21]. Experiments are conducted on magnetic field of an air core moving coil PMTLM with dual Halbach array, as shown in Fig. 6.19. The coil number of the single phase winding can be either three or five, depending on the connection of the external circuit. The main structure parameters of the linear machine are

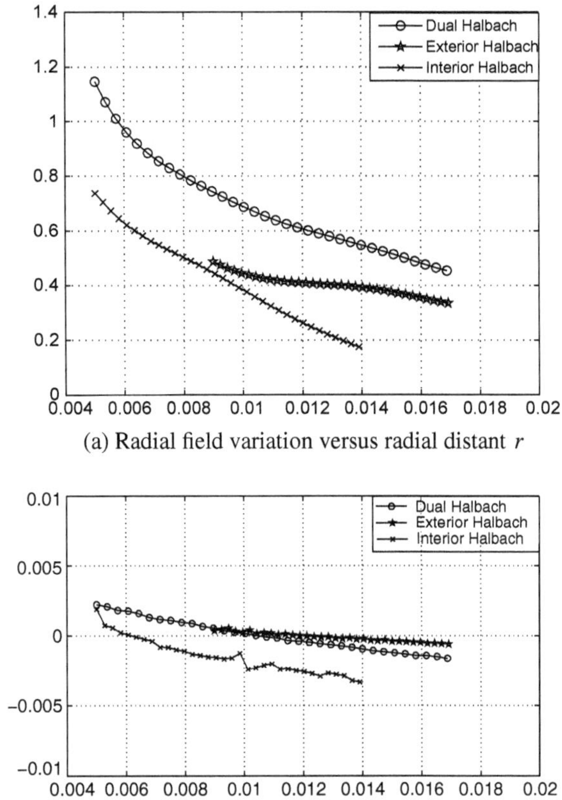

(a) Radial field variation versus radial distant r

(b) Axial field variation versus radial distant r

Fig. 6.14 Magnetic field distribution at $z = 0\,\text{mm}$

listed in Table 6.2. The current input per turn is 2 A. The comparisons between the analytical results, FEM computations, and experimental data are shown in Figs. 6.20 and 6.21 for different coil numbers. The difference between the experimental results and the analytical solutions may mainly be caused by the measurement error and the manufacturing error.

Fig. 6.15 Magnetic field distribution at $r = 12\,\text{mm}$

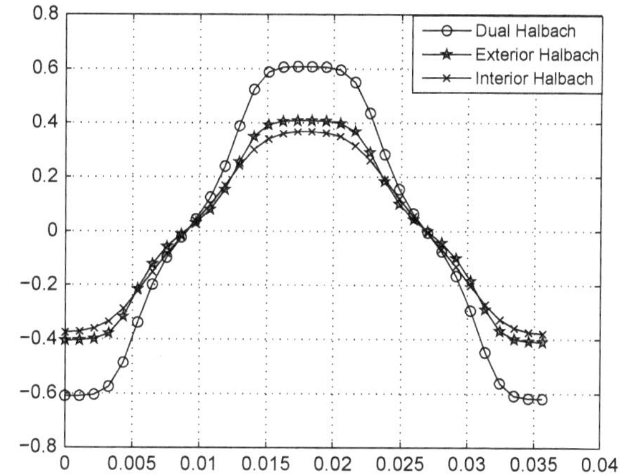

(a) Radial field variation versus axial distant z

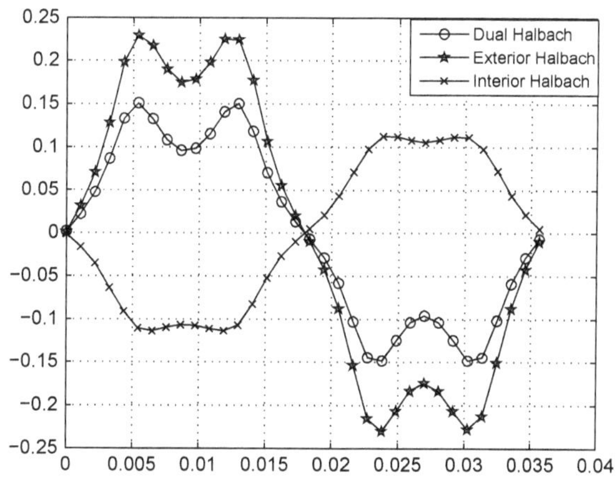

(b) Axial field variation versus axial distant r

6.5 Experimental Investigation on Inductance and Comparison

The experiments on inductance are conducted by removing resistance from the impedance as the capacitive reactance is ignorable in the winding. Therefore, two steps are carried out to measure the winding inductance. Specifically, Step 1 is to

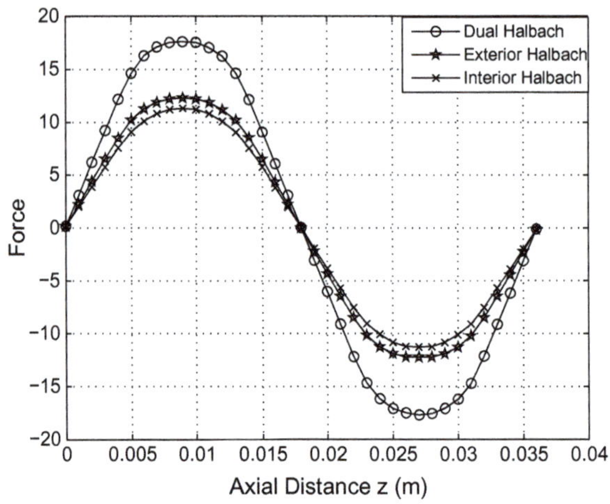

Fig. 6.16 Force comparison of different magnet arrays

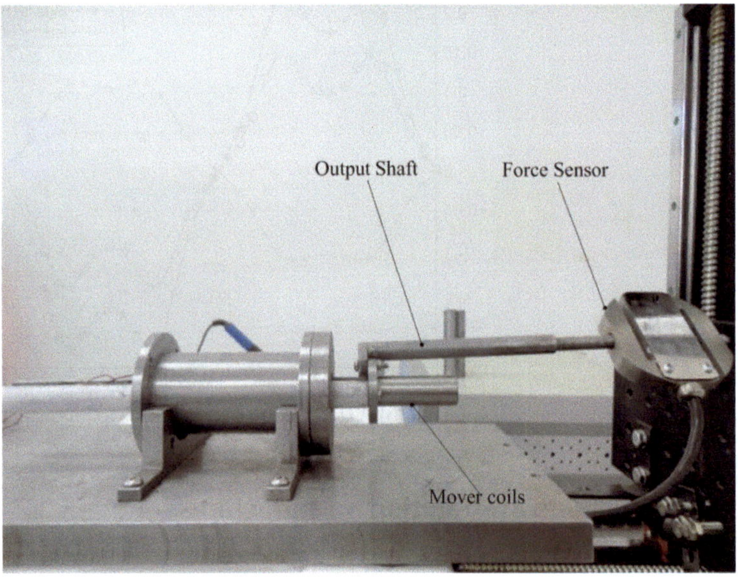

Fig. 6.17 Measurement platform of the force output

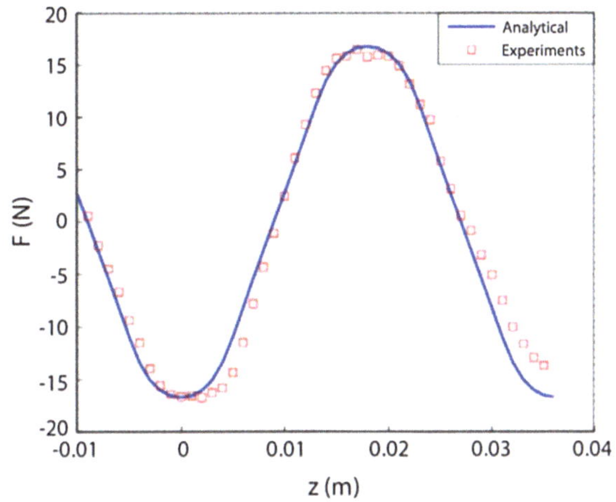

Fig. 6.18 Force output of experimental versus analytical results

measure the resistance of winding with multimeter, and Step 2 is to measure the impendence by inputting alternating current (AC) with a frequency of 50 Hz. The schematic diagram is shown in Fig. 6.22. The experimental platform is shown in Fig. 6.23. Comparison of analytical results, FEM computations, and experimental data is presented in Fig. 6.24. It is found that the three sets of data fit with each other quite well. The difference between the analytical and the measurement results is mainly caused by the measurement error and the fabrication error of windings.

However, because the self-inductance of a coil is related with its neighboring field, the self-inductances of the five coils are not exactly the same. Meanwhile, the inductive coupling coefficient between two coils is less than 1, and thus the mutual inductance is always less than the self-inductance. Therefore, Eq. (4.25), based on the assumption, leads to a relatively large difference between measurement data and analytical solutions.

Experiments are also conducted to test the linearity of the electromagnetic machine, and thus to further validate the established mathematical models. Figure 6.25 gives experimental results of maximum force output versus current input. The linear property can benefit the motion control implementation. The thrust per volume and the thrust per copper losses of different magnet topologies in given volume are investigated with FEM. The result shows that the dual Halbach array helps to increase force density of the system compared with single layer of magnet array.

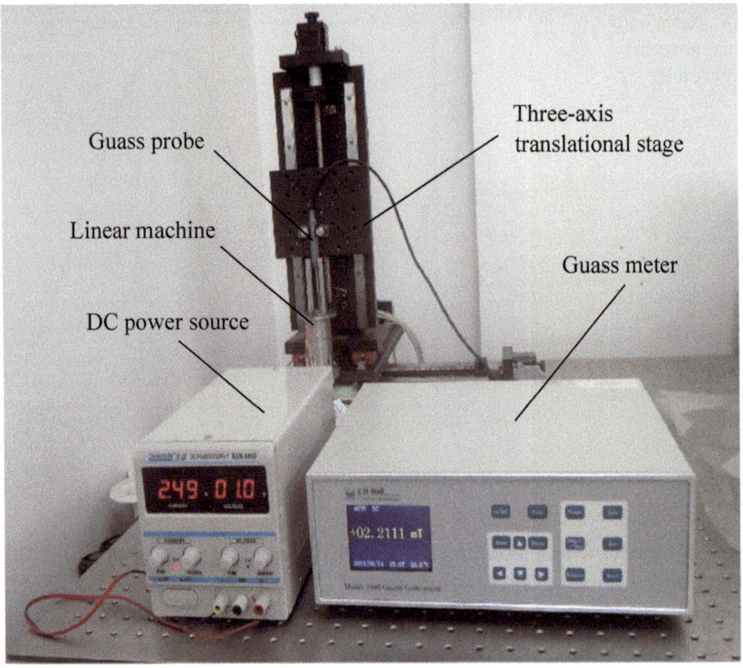

(a) Experimental apparatus

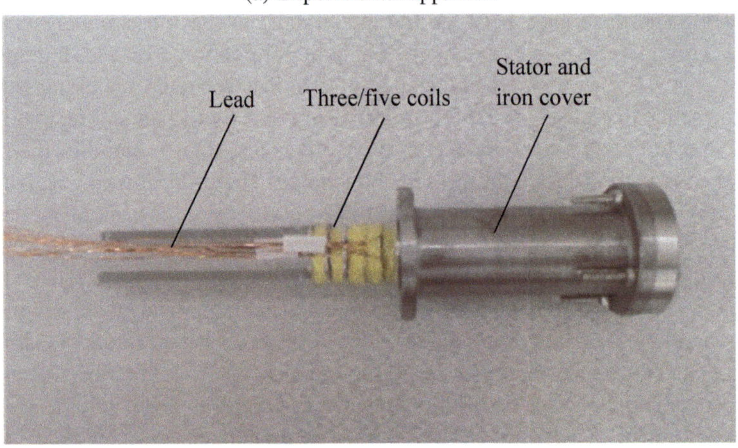

(b) Research prototype

Fig. 6.19 Experimental study on armature reaction field

Table 6.2 Structure parameters in experiments

r_i (mm)	r_a (mm)	r_b (mm)	r_o (mm)	τ_p (mm)	l (mm)	N_k
5	9.35	13.65	17	6	40.5	30

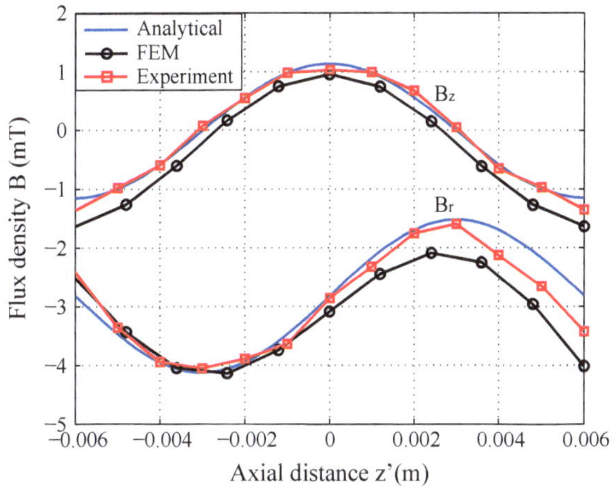

Fig. 6.20 Armature reaction field versus z' for a three-coil winding

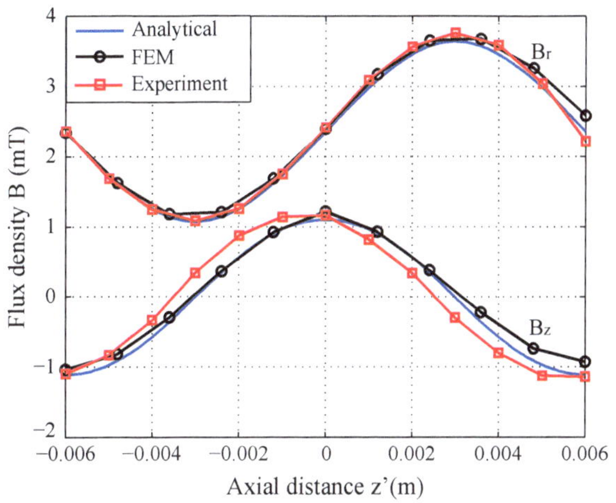

Fig. 6.21 Armature reaction field versus z' for a five-coil winding

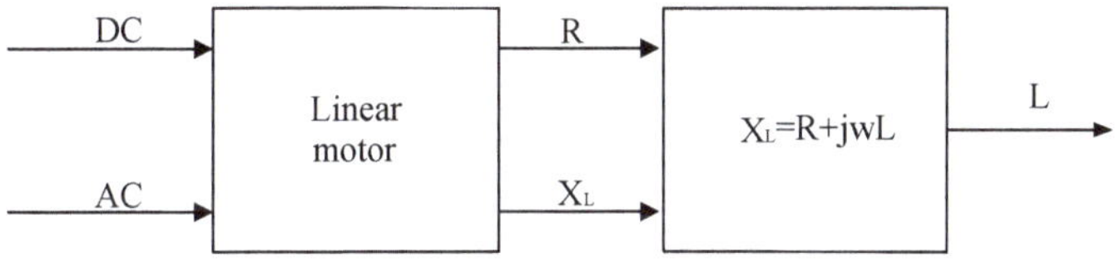

Fig. 6.22 Schematic diagram of inductance measurement

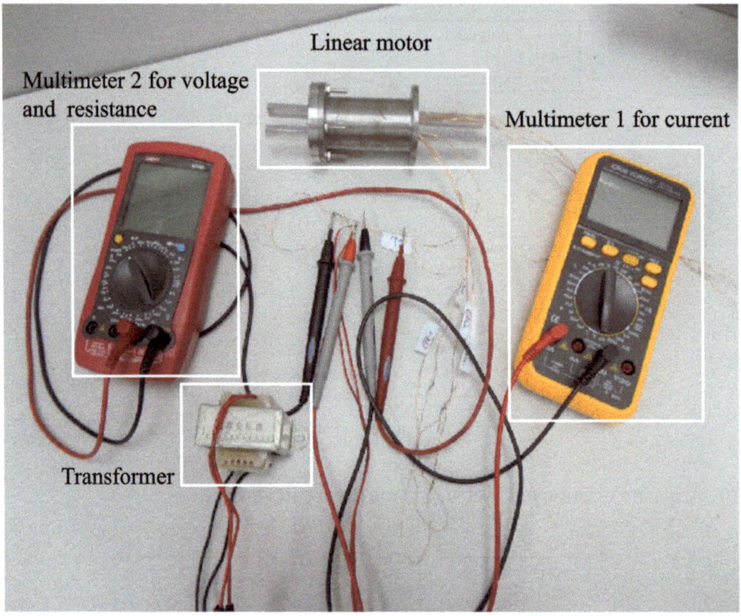

Fig. 6.23 Inductance measurement on the research prototype

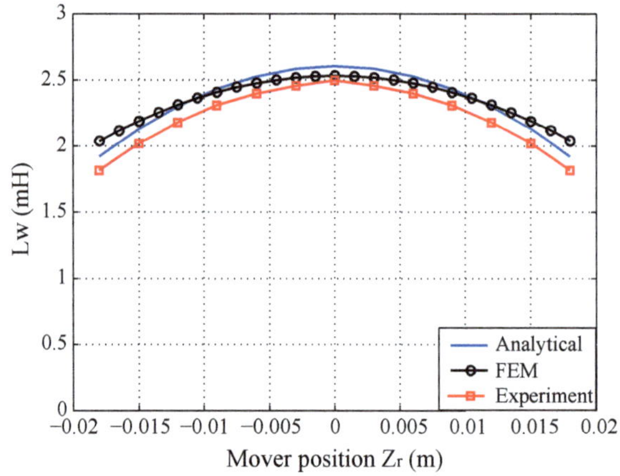

Fig. 6.24 Inductance variation versus mover position z_r validated by experiments

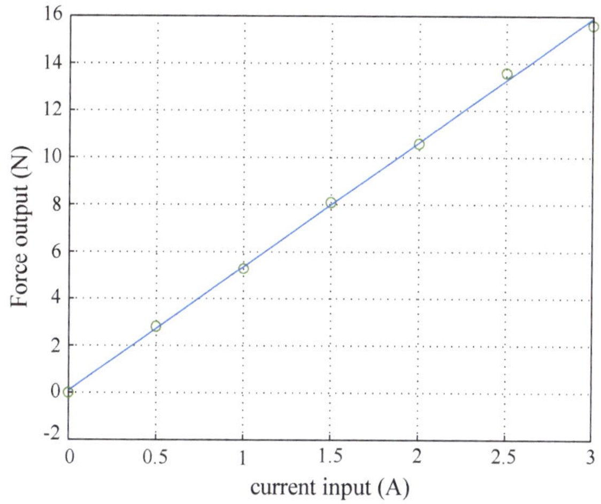

Fig. 6.25 Experimental tests on the linear relationship of force output versus current input

6.6 Conclusion

In this chapter, a research prototype with dual Halbach array, measuring apparatuses for magnetic field, force output, and inductance have been developed. Experiments have been conducted subsequently. The experimental results are compared with analytical models of magnetic field, force output, and inductance. It shows that the experimental results fit with analytical models well.

References

1. Omura M, Shimono T, Fujimoto Y (2015) Development of semicircular tubular core-less linear motor and its motion control. IEEJ Trans Ind Appl 135(3):246–257
2. Cheema MAM, Fletcher J, Xiao D et al (2015) A direct thrust control scheme for linear permanent magnet synchronous motor based on on-line duty ratio control, vol 31, no. 6, pp 4416–4428
3. Wang T, Yan L, Jiao Z (2015) Design of a novel integrated position sensor based on Hall effects for linear oscillating actuator. Rev Sci Instrum 86(7):075001
4. Tomczuk B, Sobol M (2005) A field-network model of a linear oscillating motor and its dynamics characteristics. IEEE Trans Magn 41(8):2362–2367
5. Xu F, Hu J, Zou J et al (2014) Comparative investigation of permanent magnet linear oscillatory actuators used in orbital friction vibration machine. Int J Appl Electromagn Mech 45(1–4):581–588
6. Clark RE, Smith DS, Mellor PH et al (1995) Design optimisation of moving-magnet actuators for reciprocating electro-mechanical systems. IEEE Trans Magn 31(6):3746–3748
7. Chen X, Zhu ZQ (2011) Analytical determination of optimal split ratio of E-core permanent magnet linear oscillating actuators. IEEE Trans Ind Appl 47(1):25–33
8. Chen X, Zhu ZQ, Howe D (2009) Modeling and analysis of a tubular oscillating permanent-magnet actuator. IEEE Trans Ind Appl 45(6):1961–1970

9. Wheeler PW, Clare JC, Trentin A et al (2013) An overview of the more electrical aircraft. Proc Inst Mech Eng Part G: J Aerosp Eng 227(4):578–585
10. Wang L, Zhou W, Gan Z (2011) Performance testing of linear compressors with RC approach. In: Advances in cryogenic engineering: transactions of the cryogenic engineering conference-CEC, vol 1434, no. 1, Spokane, Washington, USA, pp 1624–1631, 13–17 June 2011
11. Macione J, Nesbitt S, Pandit V et al (2012) Design and analysis of a novel mechanical loading machine for dynamic in vivo axial loading. Rev Sci Instrum 83(2):025113
12. Friedman SJ, Barwick B, Batelaan H (2005) Focused-laser interferometric position sensor. Rev Sci Instrum. 76(12):123106
13. Miklos A, Szabo Z (2015) Simulation and experimental validation of the dynamical model of a dual-rotor vibrotactor. J Sound Vib 334:98–107
14. Bajic JS, Stupar DZ, Dakic BM et al (2014) An absolute rotary position sensor based on cylindrical coordinate color space transformation. Sens Actuators A: Phys 213:27–34
15. Cheng CH, Hung SC, Liu WF (2014) A motor's rotational angle sensor based on fiber Bragg grating. Microw Opt Technol Lett 56(6):1449–1452
16. Teo TJ, Zhu H, Pang CK (2014) Modeling of a two degrees-of-freedom moving magnet linear motor for magnetically levitated positioners. IEEE Trans Magn 50(12):1–12
17. Liang H, Jiao Z, Yan L et al (2014) Design and analysis of a tubular linear oscillating motor for directly-driven EHA pump. Sens Actuators A: Phys 210:107–118
18. Leidhold R, Mutschler P (2007) Speed sensorless control of a long-stator linear synchronous motor arranged in multiple segments. IEEE Trans Industr Electron 54(6):3246–3254
19. Cupertino F, Giangrande P, Pellegrino G et al (2011) End effects in linear tubular motors and compensated position sensorless control based on pulsating voltage injection. IEEE Trans Industr Electron 58(2):494–502
20. Cheok AD, Ertugrul N (2000) High robustness and reliability of fuzzy logic based position estimation for sensorless switched reluctance motor drives. IEEE Trans Power Electron 15(2):319–334
21. Foo G, Rahman MF (2010) Sensorless sliding-mode MTPA control of an IPM synchronous motor drive using a sliding-mode observer and HF signal injection. IEEE Trans Industr Electron 57(4):1270–1278

Chapter 7
Conclusion

In view of the challenge of force improvement existing in the electromagnetic linear machine, two-dimensional magnet arrangement is expanded to three-dimensional pattern. A tubular linear machine with dual Halbach array is thus proposed to improve the magnetic flux density and the system force output. Specifically, it can not only increase the radial component of flux density that is important for axial force generation, but also decrease the radial component of force output that may cause vibrations. This monograph has presented study on the complex magnetic field in three-dimensional space, force output, and inductance. Analytical models are obtained and numerical result from finite element method is employed to validate the analytical models. Then based on the derived analytical models, design optimization has been conducted on the tubular linear machine. Structure parameters and back-iron patterns have been optimized aiming at maximizing force output. The major contributions of this investigation are summarized as follows.

- The novel design concept of tubular linear machine with dual Halbach permanent magnet array has been proposed. It helps to improve the radial flux, and reduce the axial flux. Based on the dual Halbach array, the magnetic field surrounding the mover is formulated using Laplace's equation and Poisson's equation. By utilizing the Bessel function and the harmonic expansion of the residual magnetization vector, the coefficients in the general solution of the vector potential can be determined. Because magnetic intensity is the curl of magnetic vector potential, the magnetic flux density distribution in the air space is thus obtained.
- With the knowledge of the magnetic field expression, the force modeling of the linear machine can be carried out to relate the current inputs to the force output. According to Lorentz force law, the force output caused by single-phase, double-phase, and three-phase power can be formulated.
- By taking finite machine length into consideration, an analytical modeling method to describe armature reaction field in moving-coil PMTLMs is proposed. This method absorbs the merits of Fourier analysis and permeance model. The established model of armature reaction field is validated by FEM solutions, and utilized to formulate winding inductance. It can benefit dynamic analysis and control of electric linear machines.

L. Yan et al., *Electromagnetic Linear Machines with Dual Halbach Array*, DOI 10.1007/978-981-10-2309-5_7

- Finite element method is employed to calculate the numerical results of magnetic field in three-dimensional space, force output, armature reaction field, and inductance of linear machines. The results are compared with the analytical models, which validates the obtained analytical models.
- Optimization of the tubular linear machine includes two major parts, i.e., structure parameter optimization and back-iron design. According to the proposed force model, the effect of structure parameters on the output force can be described. Thus, appropriate values for different parameters can be chosen to achieve maximum force output. Furthermore, numerical method is employed to simulate the magnetic field and force output of tubular linear machines with various type of back irons. Simulation result shows that internal back iron does help to enhance the magnetic field flux density and improve the force output, while external back iron has little effect on them. The back irons have two important functions. One is to reduce the magnetic energy loss, and the other is to enhance the magnetic flux density in the radial direction.
- A research prototype and experimental apparatus are developed for experimental purpose. Experimental investigation of the linear machine include three major parts, i.e., magnetic field distribution, force variation, and inductance. An automated apparatus has been developed to measure the flux density variation of the mover in 3D space. Because the magnetic flux density is a 3×1 vector, a three-axis Hall probe has been employed for the measurement. Measurement of force output and inductance have been conducted too. The experimental results are compared with analytical models, and validate the system operating principle.

More works could be conducted on the novel dual Halbach array such as control implementation of tubular linear machine. The employment of suitable control algorithm helps to fully take advantage of dual Halbach array. It is worth pointing out that dual Halbach array is one particular case of three-dimensional magnet array. More exploration of three-dimensional magnet arrays could be carried out to achieve high-output performance of electromagnetic machines.

Index

L. Yan et al., *Electromagnetic Linear Machines with Dual Halbach Array*, DOI 10.1007/978-981-10-2309-5

MIX
Papier aus verantwortungsvollen Quellen
Paper from responsible sources
FSC® C105338

If you have any concerns about our products,
you can contact us on
ProductSafety@springernature.com

In case Publisher is established outside the EU,
the EU authorized representative is:
Springer Nature Customer Service Center GmbH
Europaplatz 3, 69115 Heidelberg, Germany

Printed by Libri Plureos GmbH
in Hamburg, Germany